Environmental Policy and Politics

The HarperCollins Public Policy Series

Environmental Policy and Politics

Toward the Twenty-First Century

Michael E. Kraft

UNIVERSITY OF WISCONSIN–GREEN BAY

HarperCollins*CollegePublishers*

Acquisitions Editor: Leo A. W. Wiegman
Project Coordination, Text and Cover Design: York Production Services
Cover Photograph: The Picture Cube, Inc., ©Michael J. Howell
Photo Research: York Production Services
Electronic Production Manager: Mike Kemper
Manufacturing Manager: Helene G. Landers
Electronic Page Makeup: York Production Services
Printer and Binder: R.R. Donnelley and Sons Company
Cover Printer: Phoenix Color Corp.

Environmental Policy and Politics: Toward the Twenty-First Century

Library of Congress Cataloging-in-Publication Data

Kraft, Michael E.
    Environmental policy and politics
    Michael E. Kraft
        p.   cm.   (The HarperCollins public policy series)
    Includes bibliographical references (p. 209) and index.
    ISBN 0-06-501206-2
    1. Environmental policy—United States.   2. United States—Politics and govern-
ment—1993–   3. Pollution—Government   policy—United   States.   I. Title.   II.
Series
GE180.K73   1995
363.7'00973—dc20   95-17190
                                                                        CIP

95 96 97 98   9 8 7 6 5 4 3 2 1

CONTENTS

FOREWORD

This volume is the second in the HarperCollins Public Policy Series, studies of significant domestic policy issues facing the American people as we move into the twenty-first century. Designed both to complement public policy textbooks for advanced undergraduates and graduate students and to provide a resource for specialists, each of these brief works provides a comprehensive, coherent, scholarly account of the development of a particular substantive policy domain. In doing so these books offer a historical review of policy development in a given domain as well as an analysis and evaluation of current efforts to deal with key problems.

Unlike much of the writing in the field of policy studies, the books in this series are more concerned with the ideas that lie behind policy and their implementation than with issues of political process. Providing an account of the drama of congressional give-and-take in passing the Food Stamp program, for example, or of interest group strategic considerations in tax reform politics is less important here than the exploration of the sources and substance of competing definitions of problems and the resulting programs that emerge. These books, in short, immerse us more in the substance of public policy than the politics that surround its creation.

Thus, the meat of public policy in this series is conceived as a set of government objectives and programmatic initiatives that evolve over time. Rest assured this series does not and cannot ignore politics. Any public policy undergoes an evolution driven by shifting definitions of the problem which programs or government positions attempt to address, by changing political and economic considerations, and by the learning experiences gained through the implementation process. These elements, of course, are all contested in the political arena. It is clearly essential to understand the array of actors and interests involved in this competition. Furthermore, it is important to take account of the structure of public opinion, which provides a matrix in which contests over problem definitions and solutions occur. Thus, although the books in this series are principally concerned with definitions, programs, and implementation, they are at the same time attentive to the need to establish the political context in which substantive policy initiatives are developed and later modified or abandoned.

Public problems are, of course, rarely solved. Drug use, poverty, environmental degradation, and inadequate housing are perennial issues. Not only is our social technology often inadequate to eliminate social and economic problems (in contrast, say, to our ability to eliminate polio with scientific technology), but we must make choices about how to allocate our resources. There is always a question, then, of what remains to be done. Each book in this series not only looks backward, examining history as a way of understanding where we are at the present, but also forward to the policy challenges that face us at the beginning of the twenty-first century. What advances in knowledge must we make to address a particular public problem more successfully? Must we shift our definition of what the problem is? Are our approaches to problem solution evolving in ways different from the past? What changes in the political and social landscape will bear on the policy domain in question in the future? To confront these questions armed with the benefits of historical perspective is perhaps to be as well prepared for what lies ahead as we can be.

Peter Eisinger, Series Editor
La Follette Institute of Public Affairs
University of Wisconsin

PREFACE

Environmental policy is at a critical crossroads. Since the dawn of the modern environmental era in the late 1960s, public concern and the actions of environmental groups have transformed public policy. The U.S. Congress has enacted dozens of major environmental protection and natural resource policies, as have state and local governments. Environmental issues have entered the mainstream of U.S. politics, as they have in the international arena. Yet, future achievements are not guaranteed at a time when concern over the costs and burdens of implementing those policies dominate public discourse and a political backlash against environmental interests in the United States gathers momentum.

Success over the next decade depends on our ability to stimulate a constructive public dialogue on environmental policy and politics. More than ever, we need to learn what works and what does not and to employ that knowledge to redesign environmental policies for the next generation. Those policies and the political strategies used to advance them must command broad public support and be consistent with the scarce fiscal resources that constrain all public policies in the 1990s. The tasks go beyond setting budgetary priorities and fine-tuning policies dealing with clean air and water, toxic chemicals and hazardous wastes, and protection of natural resources such as wilderness areas and national parks. They involve building the political and institutional capacities for democratic governance of environmental issues in the twenty-first century.

The 1992 United Nations Conference on Environment and Development (the Earth Summit) highlighted a new array of daunting challenges that all nations face. They must learn how to respond creatively and effectively to the risks of global climate change, depletion of the ozone layer, destruction of forests and soils, loss of biological diversity, and a surging human population. No responsibility of government and society over the next generation will be more difficult than accommodating the level of economic development needed to meet human needs and aspirations while averting devastating impacts on the environment that sustains life.

Within industrialized nations like the United States, this new environmental agenda makes clear the imperative of integrating environmental protection with other social and economic activities, from energy use and transportation to agriculture and urban planning. We can no longer afford to think of environmental policy as an isolated and largely remedial activity of cleaning up the residue of society's careless and wasteful habits. Emphasis must be placed on prevention of future harm through redesign of economic activities around the concept of sustainable development.

These new needs also require that governments work closely with the private sector in partnerships that can spur technological innovations while avoiding the protracted conflicts and program delays that have constrained environmental policy achievements over much of the past 25 years. None of this means that tough regulation should be abandoned when it is necessary to achieve environmental goals. It does signal, however, the need to supplement regulation with other strategies and policy tools, such as market-based incentives and public education, that promise to speed the realization of those goals or significantly lower the costs of getting there.

Whether governments and other organizations succeed in resolving environmental problems of the 1990s and twenty-first century ultimately depends on public understanding, support, and activism. Environmental policies will require significant social and behavioral changes that will be possible only if people educate themselves about the issues and work cooperatively for environmental sustainability. Survey data confirm a high level of public concern about the environment within the United States and globally. The proliferation of thousands of grassroots organizations worldwide testifies to the public's eagerness to end environmental degradation and protect their communities. At the same time, the impact of environmental and resource policies has spurred the rise of antienvironmental organizations that have displayed considerable political influence. To be effective over the next several decades, environmentalists will need to hone their political skills and forge broader alliances with new constituencies around common interests in environmental sustainability.

This book seeks to contribute to these developments by offering a succinct overview and assessment of U.S. environmental policy and politics in the mid-1990s. It differs from other texts in several respects. It uses a risk-based framework for environmental policy analysis to encourage students to judge for themselves the significance of environmental conditions and trends and the often disputed evidence of the human and ecological risks they pose. It summarizes an extensive collection of scientific studies, government reports, and policy analyses to convey the nature of environmental problems, their causes, progress in dealing with them, and policy implications. The problems covered include global threats such as climate change, loss of biodiversity, and population growth as well as more conventional topics such as air and water pollution, toxic chemicals, hazardous wastes, and energy use.

The text also describes major U.S. environmental and natural resource policies, their origins, the key policy actors who shape them, their achievements and deficiencies, and the latest proposals for policy change. It offers a thorough, yet concise, coverage of the U.S. policy-making process, the legislative and administrative settings for policy decisions, and the role of environmental groups and public opinion in environmental politics. It gives special attention to recent debates over environmental costs, benefits, and risks, and the economic impacts of environmental policies. It covers developments in the Clinton administration through early 1995.

Chapter 1 introduces the subject of environmental policy and places it within late twentieth century political and economic trends that affect it. It also sets out an analytic framework drawn from policy analysis and environmental science that helps to connect environmental policy to the political process. Chapter 2 reviews current knowledge on the state of the environment and serves as a brief description of the environmental problems that governments and others confront in the 1990s.

Chapter 3 has two related purposes. It surveys the characteristics of U.S. government and politics and the policy-making process, with special attention to agenda setting. It then uses those concepts to trace the evolution of environmental policy and politics from the earliest days of the nation through the emergence of the modern environmental movement and reactions to it. The public's concern over the environment and its support for public policy actions are given special treatment.

The heart of the book, Chapters 4 and 5, turns to the substance of U.S. environmental and natural resource policies. Chapter 4 discusses the major environmental protection, or pollution control, statutes administered by the U.S. Environmental Protection Agency (EPA) and their continuing evolution. It also assesses the EPA, its capacity for implementing those policies, its working relationships with the states, and the processes of environmental standard setting and rule making. Chapter 5 offers comparable coverage of energy and natural resource policies. It outlines the history of efforts to design national energy policies, including the Bush administration's National Energy Strategy and Clinton administration initiatives. It describes conflicts between natural resource preservation and economic development, and puts Secretary of the Interior Bruce Babbitt's natural resource agenda within the broader context of environmentalist challenges to long-dominant natural resource interests.

Chapter 6 focuses on environmental policy evaluation. It takes a broader look at signs of progress or lack of it in achieving the goals of environmental protection and natural resource policies, and reiterates the major critiques of U.S. environmental policy in the mid-1990s. It appraises the national debate over costs and benefits of environmental policies, risk assessment, and implications of environmental policy for jobs and the economy. It also evaluates widely discussed alternatives to environmental regulation, including market incentives and privatization.

Finally, Chapter 7 looks to the future. It describes the remaining policy agenda for the 1990s, including reforms of environmental policies and institutions and issues of environmental justice. It also explores the goal of sustainable development that will shape twenty-first century environmental policy, with particular attention to policy needs in the areas of energy, population, biodiversity, environmental research, and international environmental protection. It concludes with an overview of business and the environment and the potential for democratic politics in shaping policy choices.

The book presents considerable technical material on environmental problems and public policies. I have tried to make it all as current as possible and faithful to the scientific literature. If in some instances I have fallen short of those goals, I hope readers will alert me to important omissions, misinterpretations, and other deficiencies.

In preparing a book manuscript one incurs many debts. The University of Wisconsin provided me with a year's sabbatical in 1993–1994 that was invaluable for the completion of the book. The Green Bay campus and its Department of Public and Environmental Affairs supplied me with essential office support and the technologies on which we all depend today: computers, printers, fax and copying machines, telephones, and an Internet connection. Karen Katers and Beverly Harvey assisted in innumerable ways to ensure the smooth functioning of the office that moved the project along. The organization of the book and its content have been shaped by my teaching at UW–Green Bay. I owe much to my undergraduate and graduate students whose interest in environmental policy and politics gave me the opportunity to discuss these issues at length with concerned and attentive people and to learn from them.

I have benefitted immeasurably from the individuals whose analyses of environmental politics and policy, and U.S. government and politics more broadly, make a book like this possible. Their work tells us much about how governments have responded to the environmental challenge, why they have chosen the policy approaches they have, and what difference it makes. That made my job easier. I hope the extensive references in the text adequately convey my reliance on their scholarship and my gratitude to them.

Special thanks go to my colleagues Denise Scheberle and Scott Furlong for reading the draft manuscript and counseling me in their areas of expertise. Gerrit Knaap, at the University of Illinois, was kind enough to read Chapter 6 and advise me on matters of environmental economics. Robert Bartlett of Purdue University offered many helpful suggestions on the draft manuscript. Terry Boyd and Sandra Simpson read parts of the manuscript and assisted with the index. I thank the anonymous reviewers for HarperCollins as well. They recommended a variety of changes to make the text more accessible to students. I also appreciate the splendid work of the editorial and design team at York Production Services: Tracey Topper, Candice Carta, Susan Free, and Stephanie Magean. I am particularly grateful to Norman Vig for his comments on the manuscript. This volume was stimulated and nurtured by our work in co-editing four books on environment politics, technology, and public policy. Naturally, I assume responsibility for any errors in the text that remain.

Readers will detect in these pages a tone of modest optimism often absent from works on environmental problems and public policy. It is there because I believe people have a great capacity to understand their predicament and devise solutions to it if given the right opportunities. Eric Morley, the son of my good friends Jim and Cathy Morley, inspires that confidence. So does his generation. My students encourage the same conviction, and they are smart enough to recognize the special knowledge and skills they need to work effectively on environmental problems.

This book is dedicated to the memory of my parents, Louis and Pearl Kraft. Their compassion, generosity, and caring for their fellow human beings embodied the social concerns that today are reflected in an environmental ethic. They also provided me and my brothers with the opportunity to grow up in a stunning environment in Southern California and experience the wonders of the desert, mountains, and ocean that exist in such close proximity in that part of country.

Michael E. Kraft

CHAPTER ONE

Environmental Problems and Politics in the 1990s

The approach of the twenty-first century should stimulate deep reflection on the human condition and the state of the environment in which we live. Viewed from the mid-1990s, there is abundant cause for concern. Environmental threats are more pervasive and ominous than they were at the dawn of the modern environmental movement nearly three decades ago. The problems are familiar enough: air and water pollution, toxic chemicals and hazardous wastes, destruction of ecologically critical lands and forests, loss of wilderness and wildlife, possibly severe climate changes, and poverty and deterioration of the quality of life in increasingly congested cities. Reports on these developments fill the airwaves and newspapers, and provoke public apprehension over unceasing environmental degradation. Individuals also can see the evidence for themselves in their neighborhoods, communities, and regions, and it clearly affects them. In the United States and worldwide, people believe the environment is in serious decline and they strongly support efforts to reverse the trend. Their fears are shared by many of the world's policy makers, who helped set the agenda for the historic Earth Summit, the United Nations Conference on Environment and Development held in 1992 in Río de Janeiro.

The Earth Summit was the largest international diplomatic conference ever held, attracting representatives from 179 nations (including 118 heads of state). More than 8000 journalists covered the event. In addition, representatives from more than 7000 nongovernmental organizations attended a concurrent '92 Global Forum at a nearby site in Río. In the presummit planning sessions and at the conference itself, there was a palpable sense of urgency over worsening environmental problems and their implications for economic development, especially in poor nations. There was also much evidence of public determination to deal with the problems. In a postsummit UN publication containing the conference's Agenda 21 action program, the meeting's organizer Maurice Strong spoke optimistically about what he termed "a wildfire of interest and support" that the Earth Summit had ignited throughout the world

1

and that he hoped would stimulate a global movement toward sustainable paths of development (United Nations 1993, 1). Five years earlier, the report of the World Commission on Environment and Development, *Our Common Future* (1987), had similar effects. It sold 1 million copies in 30 languages and spurred extensive policy changes in governments and the private sector (Starke 1990).

If it achieved nothing else, the Earth Summit could be called a great success for giving form and substance to the concept of sustainable development. One prominent report released a year before the summit made the need for sustainable development clear:

> Since 1900, the world's population has multiplied more than three times. Its economy has grown twentyfold. The consumption of fossil fuels has grown by a factor of 30, and industrial production by a factor of 50. Most of that growth, about four-fifths of it, has occurred since 1950. Much of it is unsustainable. Earth's basic life-supporting capital of forests, species, and soils is being depleted and its fresh waters and oceans are being degraded at an accelerating rate (MacNeill, Winsemius, and Yakushiji 1991, 3).

The authors looked back to social and economic changes in the twentieth century. What lies ahead in the twenty-first century is equally striking and worrisome. The world's 1994 population of 5.6 billion people is projected to grow to between 8 and 10 billion by 2050, with 95 percent of that growth occurring in the developing nations. To feed, house, and otherwise provide for people's needs and aspirations will tax natural resources and ecological systems throughout the world. Some estimates indicate that if current forms of development continue, the world could see a five- to tenfold increase in economic activity over the next 50 years (MacNeill, Winsemius, and Yakushiji 1991). Economic activity within the United States alone may quadruple in that same time period (National Commission on the Environment 1993, xiii).

Will the Earth be able to accommodate this kind of growth, and if so, at what human and ecological cost? How well equipped are the world's nations to respond to the needs created by such demographic, economic, and environmental trends? Will political systems prove able and willing to design and implement policies that can minimize the negative impacts of economic activities on human and ecological health and establish sustainable patterns of resource use? This book concentrates on the last of these questions, although it also addresses the first two. It is about the relationship of environmental policy to politics, and it argues that a democratic environmental politics is essential to the development of effective public policy. I focus on the U.S. federal government but give some attention to policy developments at the state and local level and to international trends affecting U.S. policy.

ENVIRONMENTAL POLICY AND POLITICS

Politics is about social choices. It concerns policy goals and tools, and the way we organize and govern ourselves. Developments of the last several decades and projections of what is coming in the next century make it clear that the United States and other nations face important decisions about their futures. They can maintain current policies and practices or they can envision a better future and design the institutions and policies necessary to help implement change. They can rely on the invisible hand of the marketplace or they can try to accelerate and consciously direct a transition to that future in other ways. Such decisions are at least as important in the United States as they are in other nations, and the ecological consequences are probably greater than for any other nation in the world. The National Commission on the Environment captured the choices well:

> If America continues down its current path, primarily reacting to environmental injuries and trying to repair them, the quality of our environment will continue to deteriorate, and eventually our economy will decline as well. If, however, our country pioneers new technologies, shifts its policies, makes bold economic changes, and embraces a new ethic of environmentally responsible behavior, it is far more likely that the coming years will bring a higher quality of life, a healthier environment, and a more vibrant economy for all Americans (1993, xi).

The commission concluded that "natural processes that support life on Earth are increasingly at risk," and as a solution it endorsed, as did nations at the Earth Summit, sustainable development. It could serve, the commission said, as a "central guiding principle for national environmental and economic policymaking," which it saw as inextricably linked, and thereby restore environmental quality, create broad-based economic progress, and brighten prospects for future generations. Consistent with its bipartisan composition, the commission observed that such a strategy would involve a combination of market forces, government regulations, and private and individual initiatives.

The nearly universal embrace of the idea of sustainable development is an important signpost for the 1990s, even if it remains a vague term that can mask serious economic and political conflicts. Such a widespread endorsement of sustainability would not have been conceivable a generation ago. In the early 1970s, books describing the "environmental crisis" and proposing ways to deal with "limits to growth" and "ecological scarcity" may have won over some college audiences and urged on nascent environmental organizations. Yet they had little discernible effect on the higher reaches of government and corporate officialdom in the United States and most other nations. In the 1990s, the language of sustainability penetrates to that level, and it has kindled grassroots activity around the world.

These developments signal fundamental changes that have occurred in U.S. and global environmental policy and politics over the past generation. The initial environmental agenda of the late 1960s and 1970s focused on air and water pollution control and preservation of natural resources such as parks and wilderness. The problems were thought to be simple and the solutions obvious and relatively easy. Public and congressional enthusiasm for environmental protection policy supported the adoption of innovative and stringent federal programs that would force offending industries to clean up even if policy goals exceeded technical and administrative capacities. In such a political climate, costs were rarely a consideration.

As the new policies were implemented in the 1970s and 1980s, their ambitious goals proved to be far more difficult to achieve than ever imagined, and much more costly. By the late 1970s, policy makers and environmentalists became increasingly frustrated with the slow pace of progress, and complaints from regulated industry and state and local governments mounted. Ronald Reagan's presidency regularly accentuated the deficiencies of the old policies (Vig and Kraft 1984). Consensus on new directions, however, was harder to achieve as Congress and the environmental community blocked what they considered ill-informed and unwise assaults on their handiwork by White House ideologues. The evidence could be seen in years of policy gridlock in the 1980s when no agreement could be reached on renewing the major environmental statutes (Kraft 1994a).

Some of the same conflicts continue in the 1990s, although with notable exceptions such as passage of the monumental Clean Air Act of 1990. In 1994 reauthorization of the leading federal policies was mired in Congress over disputes about how to balance environmental protection and economic goals. The election of Bill Clinton and Al Gore failed to green the political climate on Capitol Hill as much as environmentalists had hoped. The environmental lobby found itself stymied by intense opposition from business and industry. Even state and local governments, weary of their struggle to comply with costly federal environmental mandates, criticized proposed policy changes they considered too expensive (Camia 1994c). Environmentalists spoke openly of the possibility they might have to oppose renewal of the Clean Water Act, the Endangered Species Act, the Safe Drinking Water Act, and Superfund if they could not head off weakening amendments (Cushman 1994g).

Two months after voicing such concerns, the combined leadership of the nation's major environmental groups issued an extraordinary appeal for citizen action to turn the tide, stating that "two decades of environmental progress are in danger of being rolled back" (Environmental Coalition 1994). By the time the 103rd Congress adjourned in October, the only major environmental policy on which lawmakers could agree was the California Desert Protection Act, establishing new wilderness areas in that state. All of the other proposals would have to be revisited in the 104th Congress beginning in 1995. With Republican majorities in each house eager to fulfill the party's "Contract With America," the new Congress was far less supportive of the environmentalists' agenda.

The convening of the Earth Summit, the national commission's report, and these continuing political fights on Capitol Hill convey an important message about environmental policy and politics in the mid-1990s. Even as consensus builds for the idea of environmentally sustainable development, defining its practical meaning and formulating specific policies remain highly contentious. In this sense, one can detect continuities with the old environmental politics even as a new era unfolds that is rich in hope and possibilities.

In many ways, environmental issues in the 1990s are broader and deeper than ever before. Emerging (or reemerging) problems such as energy use, global climate change, population growth, and threats to biological diversity have extended the reach of environmentalism well beyond the old boundaries. At the same time, environmental issues, both new and old, pose deeper challenges to society as the costs and impacts of environmental policies grow, the uncertainties and complexities of the problems increase, and effective and acceptable solutions prove elusive.

The commission's report and the success of the Earth Summit symbolize the degree to which the environment has matured politically over the past 25 years and now commands broad, if not always deep, public and governmental support. As legislative fights in Congress indicate, such developments do not eliminate controversies in dealing with environmental and natural resource issues at home and abroad. There will continue to be hard-fought battles over environmental policies and their implementation. The outcomes will be unpredictable and will depend on how the issues are defined, the state of the economy, the relative influence of opposing interests, and political leadership. Nothing on the horizon indicates that political conflict over the environment is likely to disappear any time soon. Given the enormity of emerging global problems, conflict may well increase markedly over the next few decades.

For these reasons, politics and government will be indispensable to the complex process of identifying, understanding, and responding to the nation's and the world's diverse environmental problems. Policy choices are inescapably political in the sense that they seek to resolve conflicts inherent in the balancing of environmental protection and other social and economic goals. They involve a struggle over whose definition of the public interest should prevail and precisely how we should reconcile environmental goals with other competing values such as economic well-being. Most political scientists would not dispute such assertions, but in an age of rampant political cynicism, others might well doubt the government's ability to improve the environment. The role of government and politics must be assessed in the context of other leading perspectives on environmental problems and their causes and solutions.

PERSPECTIVES ON ENVIRONMENTAL PROBLEMS

Even a casual reading of the environmental literature and political commentary reveals disparate views of ecological problems and what ought to be done about them. This is not surprising. Definitions and understanding of any public problem are affected by political ideologies and values, professional training, and experience, which vary greatly across society and among scientists and policy makers. At least three major perspectives merit brief comment. These focus on science and technology, economics and incentive structures, and values and ethics.

Scientific Knowledge and Its Use

Many scientists, and business leaders as well, believe environmental problems may be traced chiefly to a lack of scientific knowledge about the dynamics of natural systems or about the impacts of technology. Or they point to a failure to put such knowledge to good use in both government and the private sector. For example, ecologists may believe that improving our knowledge of biological diversity will highlight existing and anticipated threats, and thus strengthen public policy to protect endangered species and their habitats. Better knowledge of the risks to human and ecological health of toxic chemicals would facilitate the formulation of pollution control strategies.

Adherents of this position urge increased research on environmental science, more extensive and reliable monitoring of environmental conditions and trends, better use of science (and scientists) in policy making, and the development of new technologies with fewer negative impacts. Belief in the importance of improved science and technology has prompted significant support by the federal government for research on global environmental change (more than $1 billion a year) as well as massive projects such as the $500 million National Acid Precipitation Assessment Program in the 1980s, which sought to identify and systematically describe the causes and consequences of acid precipitation, or acid rain. It also has led to increased spending on other environmental research supported by government agencies (Carnegie Commission 1992), proposals for the creation of a National Institute for the Environment, and development of policy research capabilities by major environmental organizations. For the same reasons, the business community has invested heavily in new production processes and "green" technologies (U.S. Office of Technology Assessment 1992; Underwood 1993).

Economics and Incentives

Another group of commentators, particularly economists, find the major causes of environmental ills to be less a deficiency of scientific knowledge or available technologies than an unfortunate imbalance of incentives. We misuse natural resources—especially common-pool resources such as the atmosphere, oceans, and public lands—or we fail to adopt promising new technologies such as solar power because we think we gain economically from current practice or we do not suffer an economic loss. In his classic essay on the "tragedy of the commons," Garrett Hardin (1968) illustrated how individuals may be led to exploit to the point of depletion those resources they hold in common.

We regularly see evidence of the phenomenon. In October 1994, faced with a fishery on the verge of collapse, an industry-dominated fishery management council in New England finally recommended a drastic cutback in allowable fishing in the Georges Bank area off Cape Cod. Previous limitations on fishing proved insufficient to prevent the exhaustion of the principal species that supported commercial fishing in the area for generations. Similar behavior is evident in urban areas. Individuals resist using mass transit and insist on driving their automobiles to work in congested and polluted cities even when they can plainly see the environmental degradation their behavior causes.

Because of such behavioral patterns, economists, planners, and policy analysts propose that we redesign economic and behavioral incentives that current, unrealistic market prices create (Freeman 1994; Portney 1990a; Anderson and Leal 1991). These prices send inaccurate and inappropriate signals to consumers and businesses and thus encourage behavior that may be environmentally destructive. We need, analysts say, to internalize the external costs of individual and collective decision making and establish something closer to full social cost accounting that reflects real environmental losses. This might be done, for example, with public tax policies. For instance, a steep tax on the use of gasoline and other fossil fuels would discourage their use and build demand for energy-efficient technologies. This argument extends to reform of the usual measures of economic accounting such as the Gross Domestic Product (GDP), which fail to take into consideration such losses, and even broader calls for a new paradigm of ecological economics (Costanza 1991; Daly and Cobb 1989). Only with such new price signals and revised accounting mechanisms can individual and market choices steer the nation and world in the direction of environmental sustainability.

Environmental Values and Ethics

Philosophers and kindred spirits in other disciplines, and environmentalists, offer a third perspective. The environmental crisis, they believe, is at heart a consequence of our belief systems and values, which they see as seriously deficient in the face of contemporary ecological threats whatever their other virtues may be. For example, Catton and Dunlap (1980) have defined a dominant social paradigm or worldview of western industrial societies that includes several core beliefs: humans are fundamentally different from all other species on earth over which they have dominion; people are masters of their destiny and can do whatever is necessary to achieve their goals; the world is vast and provides unlimited opportunities for humans; and human history is one of progress in which all major problems can be solved. Other premises and values follow from these beliefs: the primacy of economic well-being; the acceptability of risks associated with conventional technologies that produce wealth; a low valuation of nature; the absence of any limit to growth; support for present social and economic structures based on competition, hierarchy, and efficiency; and support for political processes characterized by centralized authority structures and bureaucratic experts (Milbrath 1989, 189). Environmentalists argue that such values strongly affect our behavior and institutional priorities and constitute one of the most fundamental causes of natural resource depletion and environmental degradation (Milbrath 1984; Leiss 1976; Nash 1989; Paehlke 1989). Even Vice President Al Gore (1992) has written about our "dysfunctional civilization" that is in disharmony with the Earth and the importance of developing an "environmentalism of the spirit" to bring about change.

Environmentalists thus contend that the values represented in the dominant social paradigm must change if human behavior is to be made more consonant with sustainable and just use of the biosphere. Aldo Leopold, one of the most celebrated advocates of an environmental ethic, wrote in his prophetic *A Sand County Almanac* that "all ethics so far evolved rest upon a single premise: that the individual is a member of a community of interdependent parts." The land ethic, he argued, simply "enlarges the boundaries of the community to include soils, waters, plants, and animals, or collectively: the land" (Leopold 1949, 239). He stated the ethic plainly: "A thing is right when it tends to preserve the integrity, stability, and beauty of the biotic community. It is wrong when it tends otherwise" (262).

Contemporary accounts reflect Leopold's perspective. Robert Paehlke, for example, derived a list of 13 values that constitute the "essential core of an environmental perspective," and he further distilled these into three key goals: (1) protection of biodiversity, ecological systems, and wilderness; (2) minimization of negative impacts on human health; and (3) establishment of sustainable patterns of resource use (Paehlke 1994). Most environmental writing espouses similar values to achieve those broad ends. These include frugality, modesty, simplicity in lifestyle, and living in harmony with nature; environmental planning based on holistic or ecological perspectives; use of "soft" or environmentally benign technologies; social and economic cooperation or a sense of fraternity; economic and political decentralization;

and participatory decision making (e.g., Ophuls and Boyan 1992; Durning 1992; Milbrath 1989).

The implication is that merely reforming environmental policies to improve short-term governmental actions is insufficient. Instead, systemic changes in political, social, and economic institutions are needed. Such arguments often are phrased in terms of the need for a new environmental paradigm or the development of an environmental ethic that can guide personal and political actions, and the role of education in bringing about such value changes. Yet there continues to be much disagreement among theorists about the specific ingredients of such a new political philosophy, the nature of the systemic changes that are needed, policy actions for the short term, and the likely effects of such changes on the environment and on government and society (Dryzek and Lester 1989; Kraft 1992; M. Lewis 1994).[1]

THE ROLE OF GOVERNMENT AND POLITICS

Each of these perspectives offers distinctive insights into both the fundamental and more immediate and direct causes of environmental problems and their solutions. All make sense though none alone offers a complete understanding or sufficient action agenda. We surely need more reliable scientific knowledge, a shift to environmentally benign technologies, more comprehensive or holistic policy analysis and planning, greater economic and personal incentives to conserve energy and other resources, and a firmer and more widely shared commitment to environmental values in our personal lives. Moreover, few would disagree that diverse actions by individuals and institutions at all levels of society and in the public as well as private sector will be essential to the long-term goal of environmental protection and sustainable development. People may choose to live close to where they work and walk or commute on bicycles or use mass transit rather than private automobiles. They may also seek durable, energy-efficient, and environmentally safe consumer goods—and consume fewer of them, recycle or compost household wastes, and avoid the common household chemicals that go down their drains and eventually into local rivers and lakes. Businesses can do much to improve energy efficiency and prevent pollution.

There remains, however, an essential role for government, politics, and public policy. Policies shape the kind of scientific research that is supported and thus the pace and character of scientific and technological developments. Similarly, governmental policies affect the design and use of economic incentives and changes in society's environmental values (e.g., through educational programs). We look to government for such policies because environmental threats represent public or collective goods problems that *cannot* be solved through private action alone. The costs may simply be too great for private initiatives and certain activities may require the legal authority or political legitimacy that only governments possess. Examples include setting aside large areas of public lands for national parks, wilderness, and wildlife preserves, and establishment of a range of international environmental, development, and population assistance programs. But much the same is true of national regulatory and taxation policies. Such policies are adopted largely because society concludes that market forces by themselves do not produce the desired outcomes. Even free market enthusiasts admit that "imperfections" in markets (such as inadequate information, lack of competitiveness, and externalities) may justify government regulation. In all these cases, private sector activity may well contribute to desired social ends. Yet the scope and magnitude of environmental problems, the level of resources needed to address them, and urgent public pressures for action may push the issues onto governmental agendas.

In these ways, the resultant public policies help to fill the gaps created when millions of individuals and thousands of corporations make independent choices in a market economy. However rational such choices are from the individual or corporate viewpoint, they are almost always guided by greater concern for personal gain and short-term profits than for the long-term social goals of a clean and safe environment. Hence, the need for establishing some limits on those choices or providing incentives to help ensure individuals and organizations make them in a socially responsible manner. Those are the preeminent purposes of environmental policy.

The goals of environmental policies and the means chosen to achieve them are set by a variety of political processes, from the local to the international level. They are also a product of the interaction of thousands of individuals and groups that participate actively in those processes. Environmentalists are well represented in decision making at most levels of government today, if not quite on a par with business and industry groups. The presence of a multiplicity of interest groups and the visibility of environmental policies usually guarantees that policy goals and instruments are subject to intense political debate, with particular scrutiny given to their cost and effectiveness. None of this diminishes the need to have such policies even as it makes clear the necessity to evaluate them carefully and to change them as the circumstances warrant.

Governmental institutions also require regular evaluation of their performance, and, where necessary, reform or more radical change. The effectiveness of government clearly cannot be taken for granted and the scarcity of public funds requires careful assessment of what works and what does not. The common standards of policy and program evaluation, such as effectiveness, economic efficiency, and accountability, are a good starting point. Students of environmental politics and policy have proposed additional criteria. These are derived from concepts of sustainability, ecological rationality, and ecosystem management (Bartlett 1986, 1990; Dryzek 1987; Ophuls and Boyan 1992). Ecological systems, or ecosystems, refer to communities of organisms living in a specific locale, the nonbiological factors (air, water, minerals) that support them, and the interactions among the components. Ecological rationality is the rationality of living systems, and it emphasizes goals and values based on the principles and concepts of the science of ecology. It would emphasize such characteristics of ecosystems as interdependence (interactive and reciprocal elements within an ecosystem or biotic community), holism (comprehensiveness and inclusiveness), and dynamic behavior (feedback, repair, and self-maintenance over time) (Bartlett 1986). This focus is consistent with recent arguments for the explicit incorporation of normative values into public policy analysis and their use in assessment of governmental programs and their effects (e.g., Anderson 1979; Stone 1988).

Much of the debate over "reinventing" or revitalizing government, made prominent in the Clinton administration's National Performance Review (1993a), is about precisely such evaluation and institutional and policy change (Osborne and Gaebler 1992; DiIulio, Garvey, and Kettl 1993).[2] At heart, the advice in such reports is not new, even if the political climate is now more supportive of serious evaluation and programmatic changes. Specific proposals for "fixing" government's environmental policies and agencies abound. They come from individual authors, environmental organizations, industry, independent policy research groups, and a wide array of governmental agencies, advisory bodies, and commissions. Many of the recommendations have entered the mainstream of political and economic thought, and they are represented on governmental agendas around the world. Policy analysts and policy makers will be debating them over the next several decades as societies struggle to adapt to a rapidly changing environment and to define and develop the most appropriate strategies and policies to navigate the uncertain territory ahead.

DEMOCRACY, POLITICS, AND ENVIRONMENTAL POLICY

In the chapters that follow, I argue that democratic decision making and public support are crucial to a successful environmental politics, and to the formulation of environmental policy that is both technically sound and politically acceptable. Democratic politics is rarely easy, and it is made especially difficult when individuals and groups hold sharply divergent perspectives on the issues and are reluctant to compromise. Protracted conflict over protection of threatened northern spotted owls and old growth forest ecosystems in the Pacific Northwest is one of the most prominent, if not especially unusual, examples of environmental gridlock. Many other cases could be cited, from nuclear waste disposal to reform of public land policies in the West. Those conditions indicate the need to establish workable political processes at all levels of government that offer hope of resolving such conflicts. There is no shortage of success stories that suggest how such processes can be developed (John 1994; Rabe 1994; Mazmanian 1992; Starke 1990).

Not all appraisals of environmental politics reach such a positive conclusion about democracy or the capacity of citizens to make the right choices. Indeed, some authors argue that democratic political systems as currently constituted are incapable of overcoming what they see as strong public resistance to the inevitable social, economic, and personal changes that will be necessary in the decades ahead. They question whether citizens will become sufficiently knowledgeable about the issues to support such demanding changes. They also are convinced that powerful economic and political interests will defeat or deflate environmental policy initiatives that threaten their privileged positions, thus preventing adequate and timely responses to environmental perils (Ophuls and Boyan 1992; Heilbroner 1991; Gurr 1985). Such beliefs lead Ophuls and Boyan, for example, to conclude that "liberal democracy as we know it—that is, our theory or 'paradigm' of politics . . . is doomed by ecological scarcity; we need a completely new political philosophy and set of political institutions" (3).

Other scholars find public concern about the environment and support for environmental protection efforts to be relatively high and persistent, if not always well grounded in knowledge of the issues (Dunlap 1992; Mitchell 1990). Several recent studies indicate that the public's ability to deal with difficult technical issues is far greater than commonly supposed (Kraft and Clary 1991; Hill 1992; Hager 1994). Still others conclude that policy makers can design public policies to promote citizen participation, enhance citizen capacities to deal with technical issues, improve political deliberation, and increase responsiveness of government to citizen needs (Berry, Portney, and Thomson 1993; Ingram and Smith 1993). Most of these scholars are more confident than the pessimistic analysts about the capacity of democratic systems to address successfully environmental and related issues. They believe that political values and institutions are malleable enough that successful change is possible with sufficient time. They are convinced, in short, that political reform works.

Whatever one's appraisal of governmental capabilities and the potential for democratic politics, there is little question that environmental policies adopted over the past three decades will have real and important effects in the United States and around the world. Some may be positive and some negative. It makes sense to try to understand their origins, current forms, achievements, and deficiencies.

DEFINING ENVIRONMENTAL POLICY

Environmental policy comprises a diversity of governmental actions that affect or attempt to affect environmental quality or the use of natural resources. It represents society's collective decision to pursue certain environmental goals and objectives and to use particular tools to achieve them, often within a specified time. We don't find environmental policy in any single decision or statute. Rather it is the aggregate of statutes, regulations, and court precedents, and the attitudes and behavior of public officials charged with making, implementing, and enforcing them. Environmental policy includes what governments choose to do to protect environmental quality and natural resources as well as what they choose *not* to do, thereby allowing other influences, such as private decision making, to determine environmental outcomes.

Policies may be tangible or largely symbolic. That is, not all environmental policies are intended to "solve" problems. Some are mainly expressive in nature. They articulate environmental values and goals that are intensely held by the public and especially by key interest groups, such as environmentalists. Such statements may have little direct relationship to legally specified policy goals and objectives, although they may nevertheless bring about important environmental changes over time by influencing public beliefs and organizational values and decision making (Bartlett 1994).

Political scientists and policy analysts have found it useful to distinguish several basic types of policies, such as regulatory and distributive policies, associated with different political patterns of policy making (Lowi 1979; Ripley and Franklin 1991). Most environmental policies fall into one category or the other, although as the case with most typologies, the fit is imperfect.

Regulatory policies attempt to reduce or expand the choices available to citizens and corporations to achieve a social goal. They may raise the cost of, prohibit, or compel certain actions through provision of sanctions and incentives. The most common approach is the setting and enforcement of standards such as the amount of pollutants that a factory or utility may emit into the air or water. Most environmental protection policies such as the Clean Air Act and the Clean Water Act are regulatory.

In contrast, most natural resources and conservation policies historically have been distributive. They have allocated or distributed public resources, often in the form of financial subsidies or comparable "particularized benefits" to clientele groups. The purpose has been to achieve nominal social goals, such as providing access to public lands for mining, grazing, forestry, or recreation; protecting biological diversity; or fostering the development of energy resources such as nuclear power (Foss 1960; McConnell 1966; Temples 1980). The U.S. Congress traditionally has favored such distributive (critics call them pork-barrel) policies that convey highly visible benefits to politically important constituencies for whom the issues are highly salient. Not surprisingly, such policies often are criticized for fostering inequitable uses of public resources and environmentally destructive practices.

A range of social, political, and economic forces shape the decisions to adopt or not to adopt environmental policies, and their impact on policy making varies, depending on the time and institutional location. As a result, the United States typically has a disparate and uncoordinated collection of environmental policies enacted at different times and for different purposes. It is rarely satisfactory from the perspective of ecological rationality or in terms of expectations for coherent and consistent public policy. As Dean Mann has noted, environmental policy "is rather a jerry-built structure in which innumerable individuals, private groups, bureaucrats, politicians, agencies, courts, political parties, and circumstances have laid down the planks, hammered the nails, plastered over the cracks, made sometimes unsightly additions and deletions, and generally defied 'holistic' or 'ecological' principles of policy design" (1986, 4).

Environmental policy also has an exceedingly broad scope. Traditionally, it was considered to involve the conservation or protection of natural resources such as public lands and waters, wilderness, and wildlife, and thus recreational opportunities and aesthetic values in addition to ecological preservation. Since the late 1960s, the term is more often used to refer to environmental protection efforts of government, such as air and water pollution control, that are grounded in concern for human health. In industrialized nations, these policies have sought to reverse trends of environmental degradation affecting the land, air, and water, and to work toward achievement of acceptable levels of environmental quality (Ingram and Mann 1983; Rosenbaum 1991).

Environmental policy extends beyond environmental protection and natural resource conservation. It includes, more often implicitly than explicitly, governmental actions affecting human health and safety, energy use, transportation, agriculture and food production, human population growth, national and international security, and the protection of vital global ecological, chemical, and geophysical systems.[3] Hence, environmental policy cuts an exceptionally wide swath and has a pervasive and growing impact on modern human affairs. It embraces both long-term and global as well as short-term and local actions.

In many respects, environmental policy resembles other public policies, as do the political processes that shape it. Yet in some important respects, environmental policy and politics are unique. There is a greater sense of urgency in addressing environmental problems, and at least some of those concerns relate to natural processes that are essential for life itself. Policy decisions also may have catastrophic or irreversible consequences that affect future generations, raising significant questions of intergenerational equity. Climate change and loss of biological diversity are good examples.

The concepts, theories, and information that contribute to environmental policy also are distinctive in their breadth and diversity. Virtually all natural and social science disciplines as well as the humanities and professional fields of study such as law and engineering play a role. Indeed, a hallmark of environmental policy is its dependence on interdisciplinary approaches for understanding human health and ecological risks and for developing technically appropriate and socially acceptable solutions. Interdisciplinary policy analysis, planning, and management are widely endorsed by students of environmental policy (World Commission on Environment and Development 1987; National Commission on the Environment 1993).

In this book I draw chiefly from political science but make considerable use of ideas and research findings from a range of disciplines, particularly policy analysis and environmental science. Political scientists focus on environmental politics rather than the content of policy itself. Yet policy and politics are closely intertwined, as the best studies of environmental policy and policy making show in areas as diverse as air pollution control (Bryner 1993), protection of biodiversity (Tobin 1990), and regulation of pesticides (Bosso 1987). The characteristics of the policy (e.g., its complexity, salience, and the impact of its benefits and costs on interest groups) determine the kind of politics that develops, such as who gets involved and who is most influential in the political process. The form of politics in turn is a strong determinant of policy outcomes.

For these reasons, it is important to study the way policy and politics interact throughout the entire policy-making process: how environmental problems emerge onto political agendas, how public attitudes and preferences affect such agendas, how policy alternatives are developed and appraised, and how policies are enacted, implemented, evaluated, and changed. It is also essential to ask about institutional and political capabilities for performing these varied public policy tasks, from the structure of the political system (e.g., the divided government and fragmented authority dictated by the U.S. Constitution) to the distribution of political power in society. Similarly, students of environmental policy need to be alert to the larger cultural, social, political, and economic contexts within which such problem-solving strategies are shaped. As Ophuls and Boyan (1992) have argued, significant constraints on environmental policy making exist in market economies and in political cultures, such as that in the United States, that accord great weight to individual rights (especially property rights) and place significant restrictions on governmental authority.

ENVIRONMENTAL PROBLEMS AND PUBLIC POLICY

Given these characteristics of environmental policy and politics, and the varied approaches to studying it, we need to ask about the nature of environmental problems and what is required to deal with them. What precisely do we mean when we say there are environmental problems, threats, or risks? Are they *public* problems—that is, unsatisfactory conditions that fail to meet public needs—or something short of that? If deemed to be public problems, under what circumstances do we look to *government* to rectify the situation? A framework derived from environmental science and policy analysis can help to clarify the questions that should be asked even if the answers depend on each individual's assessment of the problem.

In the 1990s we have been bombarded with news about a multitude of threats to public health and the environment, with varying hints about the degree of scientific consensus or the extent of disagreement between environmentalists and their opponents. Sometimes the dangers are highly visible, as they are to residents of neighborhoods near heavily polluting industries. Usually, they are not, and citizens are left confused about the severity of the problems, whom to believe, and what they ought to do. Increasingly, it seems, the so-called "third generation" of environmental problems, such as global climate change and loss of biodiversity, are even more problematic than the more familiar issues of the first generation of environmental concerns of the early 1970s (e.g., air and water pollution) and the second generation that emerged later in the 1970s (toxic chemicals and hazardous wastes). The third generation of problems are less visible, their impacts are distant and uncertain, they are often not very salient to the public, and experts disagree on the magnitude, timing, and location of their effects.

Students of environmental policy need to develop a robust capacity to sort through such partial and often biased information to make sense of the world and the public policy choices that are available. Policy makers and the public need the same skills. There is no great puzzle about the pertinent questions to ask about any public policy controversy, including those associated with the environment. They concern the scope of the problems (who or what is affected and in what ways), what to do about them, and if government intervention is called for, what kind of public policies are most appropriate. Such questions are often extremely difficult to answer with any precision, but at least making the effort enhances policy dialogue and helps to ensure that whatever solutions are pursued are rooted in valid scientific knowledge and command the support of broad segments of the public.

Defining the Problems: The Nature of Environmental Risks

One set of questions concerns the nature of environmental problems—that is, the prevailing environmental conditions and trends, their causes, and their consequences. There is a vast amount of information available on the "state of the environment," both domestic and international, and new research findings appear continually. Since its inception, the federal Council on Environmental Quality (CEQ) has produced annual reports as required by the National Environmental Policy Act of 1969, with extensive documentation of trends in U.S. air and water quality, energy use, fisheries and wildlife, production and use of hazardous chemicals, and the like (e.g., Council on Environmental Quality 1993).[4] The Environmental Protection Agency (EPA) compiles and reports on much of that information. An example is the agency's annual Toxics Release Inventory on emissions of toxic and hazardous chemicals. Other U.S. environmental and natural resource agencies assemble similar information, as do international bodies such as the United Nations Environment Programme and the World Bank.

Some environmental groups also specialize in integrating, assessing, and reporting such data. The most prominent are the Worldwatch Institute, which publishes an annual *State of the World* volume and another on *Vital Signs: The Trends That Are Shaping Our Future,* and the World Resources Institute, which releases a biennial *World Resources* report as well as an *Environmental Almanac.* These and other groups, such as the World Wildlife Fund, the Union of Concerned Scientists, and academic think tanks such as Resources for the Future, also publish independent studies of environmental problems and policies.

Unfortunately, such data can provide only a partial picture of environmental problems. The data also require careful evaluation and interpretation. The raw statistical portraits tell us little about the causes of these conditions and trends, their effects, and how much change we might expect over time as agencies enforce environmental policies and those policies begin to have some impact. Studies reporting on such data often give little information about the costs of dealing with the problems or the benefits likely to result from public policies.

Forecasting Environmental Effects   Some of the most difficult projections involve estimates of the probability of future environmental effects of societal or technological changes now underway. The better studies carefully set out the assumptions that underlie their projections and describe the forecasting methods they use. Even so, uncertainty and controversy are common, and some of it typically centers on what critics may charge are unwarranted assumptions and faulty methods. The publication of the *Global 2000 Report to the President* at the end of Jimmy Carter's administration is a good example. The report was a detailed study of "probable changes in the world's population, natural resources, and environment through the end of the century" that was to serve as "the foundation of our longer-term planning" (Council on Environmental Quality and Department of State 1980). Despite elaborate justifications offered for the forecasting methods used, the report was subject to spirited attack on methodological (and ideological) grounds (Simon and Kahn 1984). The Reagan White House also sharply criticized the study and objected to its reliance on government planning and regulation that were strongly at odds with the new administration's conservative environmental policy agenda (Kraft 1984).

A more contemporary illustration of the challenge of dealing with complex environmental problems with long-range and uncertain effects is climate change. The Intergovernmental Panel on Climate Change (IPCC), a UN-sponsored body, reflected overwhelming scientific consensus in its 1990 projection that without a sharp reduction in the use of fossil fuels, carbon dioxide levels in the atmosphere could double by the year 2100. Such an increase, in combination with other greenhouse gases, could raise global mean temperatures by three to eight degrees Fahrenheit over the next century, with devastating environmental and economic consequences (Intergovernmental Panel on Climate Change 1990). Follow-up studies by the IPCC and various committees of the National Academy of Sciences (1991) supported this basic position even if the numbers change slightly. Yet scientific and political debate continues over the validity of such projections and their policy implications. Part of the reason is that questions remain about the precise relationship of greenhouse gas concentration in the atmosphere and temperature increases, the probability that temperatures will rise by a certain amount (1, 5, or 8 degrees), by when such temperature increases will occur, and the location of certain effects such as decreased rainfall. The complex climate change computer models that scientists use to make these projections cannot yet provide all the answers (Schneider 1990; Hempel 1993).

In cases like this, scientific uncertainties commonly are exploited to strengthen particular positions on the issues. Highly vocal critics of environmentalists, from scientists such as Richard Lindzen and S. Fred Singer to conservative talk show host Rush Limbaugh, fire polemical broadsides at their adversaries on climate change issues (e.g., Lindzen 1992). From their perspective, environmentalists are guilty of "inflammatory claims," "doomsday rhetoric," and "environmental overkill" (Taubes 1993; Stevens 1993). These exchanges are intriguing but rarely enlightening, and they often confuse both the scientific and policy issues. Unfortunately, contemporary politics has created a demand for simple ideas expressed in news conference sound bites and an impatience to work through complex scientific issues and tough policy choices. These tendencies are especially great when an environmental problem first becomes visible and is judged quickly and superficially. With experience, debate, and learning, ideology may give way to more thoughtful appraisals of problems and solutions.

Assessing Risks: Social and Technical Issues   As the climate change example indicates, among other questions, we need to ask when examining environmental conditions and trends what the magnitude of risks posed to public health and the environment actually is, how well we understand such risks, and how confident we can be in data and the methods used to produce the assessments. These kinds of risks are by no means the only dimensions of environmental problems worthy of concern. A focus on risks does not begin to cover cultural, social, and moral aspects of environmental problems, or even practical matters of determining whether we should conserve land and other resources for recreational or aesthetic purposes. Consider the following description by an ecologist of old growth forests in the Pacific Northwest:

> No less important [than ecological functions], ancient forests have a transcendent aesthetic and religious value in the inner landscapes of natives and newcomers alike. Their majesty inspires comparisons with the great cathedrals. Their haunting beauty and solace attract growing numbers of Northwesterners and visitors who seek connection with a wild world that is everywhere gone or going fast. Ancient forests are a national and international resource of the highest order (Norse 1990).

Few would presume to dismiss such arguments in natural resource debates, even if they speak to us in a different language than ecological risk and economic values. We can probably agree, however, that health and ecological risks are among the most widely discussed and disputed elements of environmental policy, and hence merit careful consideration even if we acknowledge that other crucial issues involving moral, cultural, or social values deserve full examination as well.

Risk assessment is a new and evolving set of methods for estimating both human health and ecosystem risks. Controversy is endemic to use of these methods, and environmentalists have been skeptical of relying on them out of fear that the severity of the problems as they see them might be reduced when subject to formal analysis (Andrews 1994). There is equal criticism from the other side, with industry and some members of the scientific community complaining that the EPA is too conservative in its risk assessment methods and related policy decisions, e.g., labeling chemicals as carcinogenic where evidence doesn't warrant such a judgment (Belzer 1991). Despite these problems, risk assessment can be a useful, if imperfect, tool for systematic evaluation of many environmental problems. It also can help establish a formal process that brings into the open for public debate many of the otherwise hidden biases and assumptions that shape environmental policy (Carnegie Commission 1993b).

The technical community (e.g., scientists, engineers, and EPA professional staff) defines risk as a product of the probability of an event or exposure and the consequences, such as health or environmental effects, that follow. Use of the term "risk" connotes an indefinite problem, occurrence, or exposure that can be described only in terms of mathematical probabilities rather than as a certain threat. Uncertainties may be associated with estimates of either the probability or the consequences, or both. The public sees risks differently, and gives greater weight to qualities such as the degree to which risks are uncertain, uncontrollable, inequitable, involuntary, dreaded, or potentially fatal or catastrophic (Slovic 1987). On those bases, the public perceptions of risk may differ dramatically from those of government scientists, as they do in the case of nuclear power and nuclear waste disposal. Scholars describe these differences in terms of two conflicting concepts of rationality and argue that both are valid:

> "Technical rationality" is a mindset that trusts evidence and the scientific method, appeals to expertise for justification, values universality and consistency, and considers unspecifiable impacts to be irrelevant to present decision-making. "Cultural rationality," in contrast, appeals to traditional and peer groups rather than to experts, focuses on personal and family risks rather than the depersonalized, statistical approach, holds unanticipated risks to be fully relevant to near-term decision-making, and trusts process rather than evidence (Hadden 1991, 49).[5]

The differences complicate the conduct, communication, and use of risk assessments, particularly when government agencies lose public trust (Slovic 1993). Yet they do not eliminate the genuine need to provide credible risk assessments to try to get a useful fix on environmental problems and decide what needs to be done about them.

Even without such disparities in risk perception between technical experts and the public, assessing the severity of environmental problems must take place amid much uncertainty. Consider the challenge of dealing with indoor air pollutants such as radon. Radon is a naturally occurring, short-lived radioactive gas formed by the decay of uranium found in small quantities in soil and rocks. Colorless and odorless, it enters homes through walls and foundations (and sometimes in drinking water), and its decay products may be inhaled along with dust particles in the air to which they become attached, posing a risk of lung damage and cancer. In 1987 the EPA declared that radon was "the most deadly environmental hazard in the U.S.," and it has estimated it is responsible for more cancer deaths (7000 to 30,000 a year) than any other pollutant it regulates.

The best evidence of radon's effects on human health is based on high-dose exposure of uranium miners. Yet extrapolation from mines to homes is difficult. Moreover, although radon experts agree that the gas is a substantial risk at levels that are two to four times the EPA's "action level" for home exposure, studies cannot confirm a significant cause and effect relationship at the lower levels found in the typical home. But EPA and other scientists believe the indirect evidence is strong enough to take action. In a 1994 report, the EPA said that up to 2200 cancer deaths each year could be prevented through a voluntary program of reducing radon in indoor air, an effort it estimated would cost only some $700,000 per life saved. Critics question how aggressively the nation should attempt to reduce radon levels, and they estimate the ultimate cost of meeting the EPA's standard for homes and other buildings could be as high as $50 billion (Cole 1993; Leary 1994).

Coping with Environmental Risks

As the radon example shows, another set of questions concerns what, if anything, to do about the problems or risks identified. Are they serious enough to require action? If so, is governmental intervention necessary, or might we address the problem better through alternative approaches, such as private or voluntary action? Such decisions are affected by political judgments about public needs and the impacts on society and the economy. Former EPA administrator William Ruckelshaus described the constraints well: "The difficulty of converting scientific findings into political action is a function of the uncertainty of the science and the pain generated by the action" (Ruckelshaus 1990, 125). Under such conditions, some policy makers favor a prudent course of action that errs on the side of environmental protection and others are equally inclined to do little or nothing until scientific knowledge improves or political conditions change. The latter position, for example, characterized acid rain policy under President Ronald Reagan and climate change policy under President George Bush (Kraft 1994a; Vig 1994).

Consensus among scientists can speed agreement on how to respond to hazards. This was the case with the decision to phase out use of chlorofluorocarbons (CFCs) and other ozone-depleting chemicals. It is unlikely, however, that many other environmental disputes can be so easily resolved given the continuing uncertainties about health and environmental effects. A more promising alternative is to improve the way we conduct risk assessments and discuss them with the public and various stakeholders. Numerous studies show persuasively that credibility and trust are crucial ingredients in effective assessment and management of risks (Slovic 1993). They also indicate that only those approaches that involve the public and other stakeholders, provide opportunities for thorough discussion of the issues, and consider public views and concerns are likely to inspire confidence in subsequent policy actions (Dunlap, Kraft, and Rosa 1993).

What Is an Acceptable Risk?   For the public and its political representatives, a difficult issue is how to determine an "acceptable" level of risk and how to set environmental policy priorities. For environmental policy, the question is often phrased as how clean is clean enough, or how safe is safe enough in light of available technology or the costs involved in reducing or eliminating such risks and competing demands in other sectors of society (e.g., education and health care) for the resources involved. Should drinking water standards be set at a level that ensures essentially no risk of cancer from contaminants, as the Safe Drinking Water Act requires? Or should the EPA be permitted to weigh health risks and the costs of reducing those risks as long as consumers are exposed to no more than a 1 in 1 million chance of developing cancer? In 1994 Congress leaned toward the latter position. Equally important is the question of who should make such judgments and on what basis.

These judgments about acceptable risk involve chiefly political, not technical, decisions. Even when the science is firm, such decisions are difficult to make in the adversarial context in which they are imbedded. Moreover, the seemingly rational task of setting priorities may pit one community or set of interests against another when so much is at stake. That is another reason to rely on democratic political processes to make those choices, or at a minimum to provide for sufficient accountability when we delegate the decisions to bureaucratic officials.

Remediation of hazardous waste sites illustrates the dilemma of making acceptable risk decisions. Recent estimates of the costs for cleaning up the tens of thousands of sites in the United States, including heavily contaminated federal facilities such as the Hanford Nuclear Reservation in Washington state and the Rocky Flats Plant in Colorado, range from $484 billion to over $1 trillion, depending on the cleanup standard used (Russell, Colglazier, and Tonn 1992). At Hanford, among the worst sites in the nation, the federal Department of Energy (DOE) and its predecessor agencies dumped 127 million gallons of toxic liquid waste on the ground, and between 750,000 and 1 million gallons of high-level nuclear waste have leaked from single-shell storage tanks, contaminating over 200 square miles of groundwater. EPA cleanup of the average site on its National Priority List (NPL) under the Superfund program costs about $30 million and takes ten years. There is also considerable community pressure to place local sites on the NPL and to clean them to the highest standards possible. But are all these sites equally important for public health and environmental quality? Should all be cleaned up to the same standard regardless of cost and despite questions about the benefits that will result? How do we consider the needs of future generations in making such judgments so present society doesn't simply pass along hidden risks to them? Is the nation prepared to commit such massive societal resources to these ends? Historically, we've done a poor job of answering those difficult questions, but many of them were addressed in Congress in 1994 when it considered renewal of the Superfund program.

Comparing Risks and Setting Priorities   Governments have limited budgetary resources and the public has demonstrated a distinct aversion to raising taxes. Given those conditions, what are the priorities among the environmental risks we face? The federal EPA has issued several reports ranking environmental problems according to their estimated seriousness. One striking finding is that the American public worries a great deal about some environmental problems, such as hazardous waste sites and groundwater contamination, accorded a relatively low rating of severity by the EPA. The public also exhibits much less concern about other problems, such as the loss of natural habitats and biodiversity, ozone depletion, climate change, and indoor air pollution, given high ratings by EPA staff and the agency's Science Advisory Board. The U.S. Congress has tended to reflect the public's views and has set EPA priorities and budgets in a way that conflicts with the ostensible risks posed to the nation

(U.S. Environmental Protection Agency 1990a; Andrews 1994). Critics charge that such decisions promote costly and inefficient environmental policies.

Public Policy Responses

A third set of issues involves examination of policy alternatives. If governmental intervention is thought essential (often a contentious proposition in the 1990s), what public policies are most appropriate? Governments have a diversified set of tools in their policy repertoires. These include provision of information, education, research and other investments, regulation, subsidy, government purchase of goods and services, taxation, rationing, charging fees for service, and creation of public trusts. Policy analysts ask which are most suitable in a given situation, either alone or in combination with others (Schneider and Ingram 1990). On climate change, for example, many scientists have favored a program of intense research to improve our scientific knowledge in association with some immediate reduction in the use of fossil fuels. The Bush administration supported the research option but resisted doing more. It adopted a "no regrets" policy where actions would be taken against the possibility of global warming only where they could be justified on some other grounds (Vig 1994).

A lively debate has arisen in the past decade over the relative advantages and disadvantages of the widely used regulatory approach ("command and control" or "standards setting and enforcement") and such competing or supplementary devices as market-based incentives, information provision, and voluntarism (Anderson and Leal 1991; Stavins 1991). Such newer approaches have been incorporated into federal and state environmental policies, from the Clean Air Act to measures for reducing the use of toxic chemicals. In recent years, both the federal and state governments also have helped to spur technological developments through their power in the marketplace. Governments buy large quantities of certain products such as computers and motor vehicles. Even without regulation, they can alter production processes by buying only those products that meet, say, stringent energy efficiency or fuel economy standards. Analysts typically weigh policy alternatives like these according to various criteria, including technical feasibility, economic efficiency, probable effectiveness, ease of implementation, and political feasibility. None of that is particularly easy to do, but the exercise brings useful information to the table to be debated.

We need to ask as well about the distribution of environmental costs, benefits, and risks across society (and internationally and across generations) in order to address questions of equity or social justice. Numerous studies suggest that many environmental risks, such as those posed by hazardous waste sites, disproportionately affect poor and minority citizens, who are more likely than others to live in heavily industrial areas with oil refineries, chemical plants, and similar facilities or in inner cities with high levels of air pollutants (Bullard 1990, 1993; Bryant and Mohai 1992). Similarly, risks associated with anticipated climate changes are far more likely to affect poor nations than affluent, industrialized nations that can better afford to adapt to a new climate regime, even at the low to middle range of climate change scenarios (National Academy of Sciences 1991). For these reasons, among the important questions about environmental policy are who suffers the environmental risks, who pays the costs of reducing them, and who gets the benefits of public policies.

For programs already in existence, we need to ask many of the same questions about policy change, including whether programs could be made more effective through higher funding levels, better implementation, institutional reforms, and similar adjustments. Some programs may be so poorly designed or so badly implemented that they fully merit early termination. Others might be improved through careful evaluation and redesign. For example, the federal government's policy on disposal of high-level radioactive waste in geological repositories has been a notable failure to date. The program is years behind schedule, and the agency in charge, the Department of Energy, is widely criticized for poor management and insensitivity to public and state opposition to the program. Public distrust of the agency is so great that its ability to achieve its statutory task of siting a repository is in doubt (Dunlap, Kraft, and Rosa 1993). Critics have called for a moratorium on implementation and a complete overhaul of the policy (Shrader-Frechette 1993).

The EPA itself has been the object of considerable criticism on many fronts, including the inadequacy of its scientific research, its inability to supervise outside contractors properly, and varied management deficiencies (Rosenbaum 1994; Landy, Roberts, and Thomas 1990). Yet it is also clear that the EPA would have achieved far more of the statutory objectives given to it by Congress had it been provided with sufficient levels of funding over the years. Much the same argument can be made for state environmental programs, which have been adversely affected for years by federal mandates unaccompanied by sufficient funding to meet them.

CONCLUSION

A simple idea lies behind all of these questions about the scope of environmental problems and strategies for dealing with them. Improving society's response, particularly in the form of environmental policy, requires careful appraisal of the problems and a serious effort to determine what kinds of solutions hold the greatest promise. We will never have enough information to be certain of risks posed, all the impacts of current policies and other activities, and the likely effectiveness of proposed courses of action. But as citizens we can better deal with the welter of data and arguments by focusing on the core issues and acknowledging that disagreements are at least as much about different political values and policy goals as they are about how to interpret limited and ambiguous scientific information. If citizens demand that policy makers do the same, the nation can move closer to a defensible set of environmental policies that offer genuine promise for addressing the problems faced.

Chapter 2 will focus on an overview of contemporary environmental problems, from pollution control and energy use to biological diversity and population growth. My intention is to ask the kinds of questions posed above related to each of the major areas of environmental concern and thus to lay out the scope of environmental problems, the progress being made to date, and the challenges that lie ahead.

ENDNOTES

1. Martin Lewis (1994) describes his book as "deliberately confrontational" and intended to provoke a reconsideration of the basic tenets of the environmental movement, particularly that branch of it populated by "eco-radicals." He urges a hard examination of its "shaky foundations" that have been taken largely on faith and out of antipathy to contemporary economic, social, and political institutions. One can disagree with much of Lewis's assessment and still argue that critical thinking about environmental goals and values and, where possible, empirical study of the effectiveness of the strategies being advocated are essential.

2. The National Performance Review (1993b) released a long series of what it called "accompanying reports," the findings and recommendations of which are discussed selectively throughout the book. They pertain to environmental regulation, natural resources policy, energy policy, and many of the federal government agencies responsible for those policies.

3. An unusual instance of Congress explicitly connecting environmental and transportation policy is the Intermodal Surface Transportation Efficiency Act of 1991 (ISTEA). ISTEA authorized $151 billion over six years for transportation that included $31 billion for mass transit. It required statewide and metropolitan long-term transportation planning, and specifically authorized states and communities to use transportation funds for public transit that reduces air pollution and energy use consistent with the Clean Air Act of 1990.

4. President Clinton initially proposed elimination of the council as part of a move to consolidate environmental functions in a cabinet-level Department of the Environment. However, Congress has been unable to approve an EPA cabinet bill, and key lawmakers persuaded the president to drop his plan to delete the council. Clinton then recommended merging his new White House Office of Environmental Policy into the CEQ. In 1994, he nominated Kathleen McGinty, director of the policy office, for the post of council chair.

5. One of the most thorough reviews of these disputes over the meaning of "rationality" in risk assessment and management and the implications for citizen participation in environmental policy is by the philosopher K. S. Shrader-Frechette (1991). See also Kraft (1994c and 1995b) and Slovic (1993).

CHAPTER TWO

Judging the State of
the Environment

If getting a firm handle on the nature of environmental problems seems inordinately complicated, that's because it is. There is no easy way to evaluate the entire ensemble of environmental threats, as the questions set out at the end of Chapter 1 indicate. Nonetheless, to judge how well the U.S. political system (and others) have responded to these problems and how well they might do in the future requires at least a brief tour of the terrain. The rest of this book reviews the development of environmental policy in the United States, describes the configuration of current policy, assesses its strengths and weaknesses, and looks ahead to a policy agenda for the twenty-first century. This chapter focuses on prevailing views of environmental problems and the evolution of society's understanding of them.

No definitive list of environmental problems can guide such a review, and as discussed in Chapter 1, environmental quality is tied inextricably to other social activities as diverse as agriculture, transportation, and energy use. Yet convention dictates that such a survey of environmental problems should highlight conditions and trends in certain areas. Among the most common are air and water quality, toxic chemicals, hazardous wastes, and solid waste, all of which I cover here. To these I add an overview of energy use and its implications for climate change, biological diversity and habitat loss, and human population growth. For each topic, I try to summarize the most important features of the problem in the United States—and where appropriate, internationally—what selective data indicate regarding the scope and severity of the problem, and what such data say about the effects of current policies or the promise of policy initiatives and recommendations. As noted in Chapter 1, we need to engage some difficult questions concerning what we know about environmental problems and their causes, the uncertainties that exist, and the actions that are necessary. There are few definitive answers and many remaining questions.

23

AIR QUALITY

Air pollution results from myriad and complex causes. It is in large part, however, attributable to the combustion of fossil fuels for power generation and domestic heating and their use in industry and motor vehicles. Other major contributing causes are the production and use of volatile organic compounds (VOCs) such as solvents in chemical plants, refineries, and commercial establishments. Urban smog occurs when nitrogen oxides from the burning of fuel and VOCs interact with sunlight to form ground-level ozone (to be distinguished from the protective stratospheric ozone layer) and other chemicals. The use of motor vehicles powered by internal combustion engines is the main source of urban air pollution in most industrialized nations. Aside from problems of reduced visibility and mild irritation, air pollution exacts a heavy toll on public health, causing thousands of premature deaths each year in the United States (Chivian et al. 1993).[1]

Gains in Air Quality and Remaining Problems

By most measures, air quality in the nation has improved significantly since the adoption of the original Clean Air Act Amendments in 1970. The environmental trend data, while not always reliable because of gaps and other deficiencies in monitoring, are encouraging. The EPA reports that for the ten-year period 1983 to 1992, air quality levels rose by 34 percent, total emissions declined by 25 percent, and highway vehicle emissions declined by 30 percent. Remarkably, these gains were achieved in the face of a 37 percent increase in vehicle miles traveled. For the same period, the nation experienced an 89 percent reduction in ambient levels of lead (primarily due to removal of lead from gasoline). Atmospheric concentrations of carbon monoxide fell by 34 percent, nitrogen oxides by 8 percent, ozone by 21 percent, and sulfur dioxide by 23 percent. Fine particulates (dust and soot less than 10 microns in diameter, known as PM-10) dropped by 17 percent since the EPA adopted new standards in 1987. Especially steep declines occurred in the last year of the study, from 1991 to 1992 (U.S. Environmental Protection Agency 1993). Paul Portney (1990b), focusing on the period 1978 through 1987, found equally impressive declines in lead, carbon monoxide, and ozone, and greater drops in particulates (21 percent), sulfur dioxide (35 percent), and nitrogen dioxide (12 percent).

These achievements from 1978 through 1992 may seem modest until one considers that between 1970 and 1990, the U.S. population increased by 25 percent and economic activity grew by about 70 percent, adjusted for inflation. Some of the declines in air pollution might have occurred even without the push of public policy. But the Clean Air Act (CAA) itself has produced significant improvements in air quality and thus in public health.

Such gains notwithstanding, the EPA found that in 1992, 54 million people lived in counties that failed to meet at least one of the national air quality standards for the six major pollutants covered by the CAA (though the number was down from 86 million in 1991). Air quality remained unacceptable in the large urban clusters in which many Americans live. An estimated 44.6 million people (down from 70 million in 1991) resided in counties where pollution levels during 1992 exceeded federal standards for ozone, the chief ingredient of urban smog. Such figures vary from year to year, reflecting changing economic activity and weather patterns (U.S. Environmental Protection Agency 1993). As the number of cars on the road and the miles driven increase, and as population rises, many areas will find it difficult to meet new federal air quality standards without taking some action affecting the use of vehicles.

Ground-level ozone in particular is a major public health threat, capable of causing respiratory difficulties in sensitive individuals, reduced lung function, and eye irritation; it can also inflict damage on forests and agricultural crops. Other major air pollutants are associated with a range of health and environmental effects, from eye and throat irritation and respiratory illness to cardiovascular and nervous system damage. Studies reported in 1993 indicate that fine particulate matter in air pollution from industrial plants could be responsible for as many as 50,000 to 60,000 deaths a year, mostly among children with respiratory problems, people with asthma, and the elderly (Hilts 1994). If confirmed, this would represent a far larger number than attributable to other forms of pollution. According to the EPA, some 25 million people in the United States live in counties that violate federal standards for fine particulates (U.S. Environmental Protection Agency 1993). The American Lung Association says the number rises to 92 million if the stricter California standard is used (Hilts 1994). Very little of the $35 billion the nation spends each year on air pollution control is directed at such small particles, though the costs of additional controls on fine particulates could be modest.

Air pollution affects even areas well removed from the nation's major urban centers. In mid-1994, visitors to many of the most popular national parks were being warned of health-threatening levels of ozone and other pollutants. Affected parks included Sequoia National Park, the Grand Canyon National Park, Mount Rainier National Park, the Great Smoky Mountains National Park, and Rocky Mountain National Park, among others. In Rocky Mountain National Park, ozone levels have exceeded 80 parts per billion more than 120 times over the past six years. At that level, ozone can cause headaches, chest pain, and labored breathing.

In addition to the six major pollutants defined in the CAA, toxic or hazardous air emissions, such as vinyl chloride, mercury, arsenic, and benzene, are of increasing concern, and are associated with a wider array of health problems, from neurological damage and birth defects to cancer. The EPA reported that some 2 billion pounds of such air toxics were released in 1991. The 1990 CAA included strong new provisions for reducing hazardous air pollutants.

The Case of Southern California

The effects of such air pollutants vary widely around the nation. They are especially severe in Southern California. In the late 1980s, for example, the South Coast Air Basin (the Los Angeles metropolitan area and adjoining counties) violated federal health standards for ozone on 137 days per year on average, with ozone concentrations sometimes triple the maximum permissible federal level. A high concentration of people and vehicles in what was already the nation's largest manufacturing center indicated the problem would not likely disappear. The basin's population of 12 million people was projected to rise to more than 16 million by 2010, its 8 million cars were expected to increase at an equally high rate, and the number of vehicle miles traveled were projected to grow by 68 percent. The basin could not be brought into compliance with federal air quality standards without drastic action.

In 1989 the South Coast Air Quality Management District adopted the nation's most ambitious air pollution control plan, to be implemented in stages over a 20-year period (South Coast Air Quality Management District 1989). It was revised in 1991 and again in 1994. In addition to extending current restrictions on auto emissions, power plants, and industry, the plan targeted smaller businesses such as furniture manufacturers, dry cleaners, and bakeries, and individual use of gasoline-powered lawn mowers, barbecue lighter fluid, and other consumer products. It also provided monetary incentives for car pooling and improving public transit.[2]

The California Air Resources Board adopted demanding air quality policies as well, including a requirement that 2 percent of cars offered for sale in the state in the 1998 model year be "zero emission" or electric vehicles, rising to 5 percent in 2001 and 10 percent in 2003. Twelve northeastern states and the District of Columbia have indicated they wish to adopt the California standards for cleaner vehicles as well as zero emission cars, helping to create a large market for a new generation of automobiles. The U.S. auto industry has firmly resisted efforts to force it to build cleaner cars, arguing that they will be too costly and fail to meet consumer expectations for performance. In 1994 the EPA was mediating between the industry and the northeastern states to try to resolve those conflicts. In recognition that insufficient progress was being made nationwide on urban air quality and toxic air pollutants, the CAA of 1990 included some of the same kinds of measures being tried in Southern California.

Indoor Air Quality

The quality of indoor air poses as high a risk for many people as does the air outside though it was ignored for years by the public and government agencies. Recently, the EPA has expressed concern about indoor pollutants, ranking them as one of the top risks to public health. Radon and environmental tobacco smoke (ETS) are especially worrisome. In January 1993, after years of extensive investigation, two public reviews, and recommendations from its Science Advisory Board (SAB), the EPA classified ETS as a known human (Group A) carcinogen and a "serious and substantial public health threat." The agency estimated that as many as 3000 lung cancer deaths in nonsmokers a year in the United States were associated with such second-hand smoke in addition to other effects such as increased incidence of bronchitis and pneumonia in young people and of asthma attacks. The report also found that ETS has subtle though significant effects on the respiratory health of adult nonsmokers, such as reduced lung function (Browner 1993).[3] Recent studies suggest ETS is a factor in heart disease as well, by some estimates contributing to as many as 30,000 deaths a year (Stone 1994a). Smokers themselves assume the greatest risk. The federal Centers for Disease Control and Prevention reports that smoking is the single most preventable cause of premature death in the United States, accounting for more than 400,000 deaths a year.

For several reasons, indoor air quality has worsened over the past several decades. Modern homes that are well-sealed to improve their energy efficiency have the distinct drawback of allowing a multitude of pollutants to build up indoors if the home is not properly ventilated. Use of synthetic chemicals in building materials and furnishings also contributes to the problem, as does use of many household cleaning and personal care products. Important indoor pollutants in addition to tobacco smoke and radon include vinyl chloride, formaldehyde, asbestos, benzene, fine particulates, lead, combustion products such as carbon monoxide from gas stoves and inadequately vented furnaces, and biological agents (Samet and Spengler 1991). Some estimates put the concentration of several toxic pollutants as high as 100 times greater indoors than they are outside, although for obvious reasons, scientific data on exposures and risks are limited. Since most modern office buildings come equipped with permanently sealed windows, the quality of indoor air that many people breathe all day depends on the workings of fine-tuned ventilating systems that sometimes fail. In some cases, the results may be acute health effects often described as Building Related Illness.

The EPA itself has struggled with a companion problem, the so-called Sick Building Syndrome, that afflicted its original Washington headquarters at Waterside Mall. Here the concern is chronic health effects associated with building characteristics. After staff complaints at the EPA mounted in the mid-1980s, the agency searched unsuccessfully for the causes of its air quality problem. Eventually a number of EPA employees had to relocate to other buildings; five of them successively sued the EPA, with a jury in late 1993 awarding nearly $1 million in damages.

In 1994 Congress was considering several bills on indoor air quality. These included measures to prohibit smoking in buildings accessible to the public except where separately ventilated smoking rooms are available, the use of market-based incentives to inform home buyers of radon risks, and the adoption of voluntary guidelines for identification and elimination or prevention of other indoor pollutants. In July 1993 the EPA announced voluntary guidelines of this kind for control of ETS in public buildings. Going even further, in 1994, the Occupational Safety and Health Administration (OSHA) was considering a proposal to require that all employers in the nation either ban smoking or provide a designated smoking area that is enclosed and ventilated directly to the outside. Such a rule would affect factories, offices, bars, restaurants, and indoor sports arenas, among other facilities. Since OSHA takes about ten years on average to approve a proposed rule, other measures are much more likely to take effect first.

Acid Precipitation

Two of the major air pollutants, sulfur dioxide and nitrogen dioxide, also contribute to acid precipitation, or acid rain. Although scientific evidence is not conclusive on the severity of all impacts, the ten-year, comprehensive National Acid Precipitation Assessment Program (NAPAP) and other studies have found that acid precipitation adversely affects aquatic ecosystems, forests, crops, and buildings. It may also threaten the health of individuals with respiratory problems and degrade visibility (National Acid Precipitation Assessment Program 1990). As a result of such concerns, the 1990 CAA required that sulfur dioxide emissions be cut by 50 percent, or by 10 million tons a year, and nitrogen dioxide by 2 million tons a year. The act established an innovative program of economic incentives through emissions trading to help meet those goals (Bryner 1993).

Regulatory initiatives for attaining national air quality standards and for reducing hazardous air pollutants and precursors to acid precipitation are costly. They impose substantial burdens on state and local governments as well as on industry, and provoke extended debate over the severity of air pollution risks and the benefits of policy action. Estimates of costs vary widely, however, and new cost-benefit studies to be completed by the EPA are unlikely to end the debate (Bryner 1993).

CFCs and the Stratospheric Ozone Layer

Some air emissions are more important for global atmospheric conditions than for urban air quality. With several other chemicals, chlorofluorocarbons (CFCs) have been linked to depletion of the stratospheric ozone layer, which shields life on earth from dangerous ultraviolet radiation. Chemically stable and unreactive, CFCs rise in the Earth's atmosphere to a level above 25 kilometers (about 15 miles) where they are broken down by intense ultraviolet light. The freed chlorine atoms can destroy as many as 100,000 ozone molecules before being inactivated. In this way, CFCs reduce the capacity of the ozone layer to block ultraviolet radiation from penetrating to the surface of the Earth. CFCs are also a major greenhouse gas, contributing to possible climate change.

Evidence of the depletion of the ozone layer has been building for the past decade, and it has been particularly notable at the North and South poles. Some estimates suggest that for every 1 percent decrease in stratospheric ozone concentrations, there is a 2 percent increase in ultraviolet B (UV-B) radiation reaching the surface of the Earth. Until recently, however, such an increase could not be confirmed outside of Antarctica. The first persuasive scientific evidence of surface increases in UV-B radiation in temperate zones was published in late 1993. It showed wintertime levels of UV-B radiation in the test area (Toronto, Canada) increased 5 percent every year between 1989 and 1993 as stratospheric ozone levels declined (Kerr and McElroy 1993). Critics continue to maintain that the ozone threat may be exaggerated, and only further scientific study can confirm these findings. Should the relationship between deterioration of the ozone layer and increased UV-B radiation hold true, the EPA has projected that a 1 percent decline in stratospheric ozone could produce a 5 percent increase in cases of nonmalignant skin cancer. It could also raise by some 2 percent the number of cases of malignant skin cancer, or melanoma, which currently kills about 5000 Americans a year. If the ozone layer depletion continues, the EPA has estimated that over the next century as many as 150 million additional cases of skin cancer and perhaps more than three million additional cancer deaths could occur. Other possible health effects include depression of immune systems, which would allow otherwise minor infections to worsen, and increased incidence of cataracts, a clouding of the eye's lens that is currently the leading cause of blindness in the world (Benedick 1991, 21)

An increase in ultraviolet radiation could also adversely affect animal and plant life, and therefore agricultural productivity. Scientists have expressed concern that higher UV radiation could severely harm oceanic phytoplankton, and thus affect the food chain as well as the capacity of oceans to regulate climate. Some of these effects have been detected in the Antarctic since 1987 (Schneider 1991).

Scientific consensus on the risks of CFC use was great enough for the industrialized nations to agree on the phaseout of CFCs, halons, and other ozone-destroying chemicals on a fairly aggressive schedule. That agreement was embodied in the Montreal Protocol, approved in 1987 after several years of international meetings. Policy makers strengthened the agreement in 1990 and again in 1992 when evidence suggested ozone depletion was more extensive than had been believed; the phaseout was accelerated to 1995. The U.S. Congress also included a supplement to the Montreal Protocol in the 1990 CAA amendments. The United States committed itself to a more rapid phaseout of the production and use of ozone-depleting chemicals and other measures such as recycling and disposal of such chemicals from discarded appliances. By late 1993 there was already evidence that these policies were having an effect; the buildup of ozone-depleting chemicals in the atmosphere had begun to slow. The chemical industry, well aware of the burgeoning new market for non-CFC refrigerants and other uses, has developed substitutes for CFCs, and it continues to search for better products. Because of the way it integrates continuous assessment of environmental trends with policy action, the Montreal Protocol is often cited as a model of global environmental governance.

WATER QUALITY

Water resources and quality are vital to life and to the nation's economy. Water resources support agriculture, industry, electric power, recreation, navigation, and fisheries, and they are distributed around the nation unevenly. Natural hydrologic conditions and cycles (particularly the amount of rain and snow) determine the amount of water in any given location. Its quality is affected by human uses, including so-called point discharges from industry and municipalities and nonpoint sources such as agriculture and runoff from urban areas.

The state of the nation's water quality is more difficult to measure than its air quality, in part because of the very large number of bodies of water and great variability nationwide in their condition. Hence evidence of progress over the past 25 years is more limited and mixed. Most assessments deal separately with surface water quality (streams, rivers, lakes, and ponds), groundwater, drinking water quality, and the quantity of water resources available for human uses such as drinking and agriculture. Both human and ecosystem health effects of water quality have been objects of concern. I will review only a few key indicators of conditions and trends here.

Pollution of Surface Waters

Despite expenditures of over $500 billion since adoption of the Clean Water Act of 1972, mostly on "end-of-pipe" controls on municipal and industrial discharges, the state of the nation's water quality is improving only slowly. Lack of reliable data make firm conclusions about the pace of progress difficult. But many studies and monitoring programs are now underway that will eventually provide a fuller accounting (Knopman and Smith 1993). Some trends, however, are fairly clear. There has been a major reduction in the raw pollution of surface waters. The percent of the U.S. population served by wastewater treatment plants rose from 42 percent in 1970 to 74 percent in 1985, with a resulting estimated decline in annual releases of organic wastes of about 46 percent. There have also been striking declines since 1972 in discharge of priority toxic organic pollutants (99 percent, according to the EPA) and toxic metals in some 22 industries. Reductions in toxic metals are estimated to be about 98 percent, or 1.6 million pounds per day (Adler 1993). Studies of long-term trends in Great Lakes water quality, an important indicator of surface water conditions, also show improvement. The concentration of polychlorinated biphenyls (PCBs) in Lake Michigan trout have fallen almost 90 percent since 1973, and the levels of DDT in the milk of nursing mothers fell from 150 parts per billion in 1967 to under 15 parts per billion in 1986. Populations of some species severely affected by pollution in the 1970s, such as bald eagles and double-crested cormorants, have climbed sharply (Schneider 1994b).

Even with such gains, many water quality problems remain. Information on water quality is published by the EPA in the agency's biennial National Water Quality Inventory, which primarily consists of data from the states. The states are able to evaluate only about 20 percent of rivers and streams and about half of the nation's lakes, ponds, and reservoirs, so the picture is incomplete if nevertheless indicative of remaining problems. In the 1994 inventory (covering the years 1990–1991), the states reported that roughly 40 percent of rivers and lakes and one-third of the estuaries assessed are not meeting or fully meeting designated uses such as swimming, fishing, drinking water supply, and support of aquatic wildlife (Adler 1994). Only 2 percent of Great Lakes waters, which contain one-fifth of the world's fresh water supply, were deemed to fully support such uses. Ninety-seven percent were rated by the states as fair or poor because of pollution from toxic chemicals such as PCBs and the pesticide DDT.

According to this database, by far the biggest source of water quality problems in rivers and streams (72 percent according to EPA) comes from agriculture in the form of nutrients (farm fertilizers and animal wastes), pesticides, and suspended solids. Urban runoff from rain and melting snow is also significant at about 15 percent. Such runoff carries a wide assortment of chemicals into local rivers, bays, and lakes or into the groundwater. Sewage treatment plants are the second largest contributor, with other causes of impaired water quality including habitat modification (e.g., loss of wetlands), storm sewers, and industry (Council on Environmental Quality 1993).

Continued loss of wetlands is particularly striking because of the important role they play in maintaining water quality. In every state except Alaska and Hawaii, over one-half of the original wetlands have been lost. In California and Ohio, the most extreme cases, the figure is closer to 90 percent. Between the mid-1970s and the mid-1980s, the nation lost an estimated 290,000 acres a year of marshes, swamps, and other ecologically important wetlands to development, highways, and mining. Only about 4 percent of the remaining 277 million acres have been assessed, but analysts found one-half of those studied unable to support all the expected human and natural uses, primarily due to sediment from agriculture and development (C. Anderson 1994; Kusler, Mitsch, and Larson 1994).

Regulatory water quality programs have concentrated on conventional sources of pollution such as biological waste products that can be assimilated and eventually cleaned by well-oxygenated water. Regulators are only beginning to deal with the more challenging problem of toxic chemicals that enter the nation's waters and their ecological effects. In some areas of the country, the impact of both conventional pollution and toxic contaminants on aquatic ecosystems has been severe. For example, Lakes Michigan, Huron, Erie, and Ontario receive up to 5000 pounds of PCBs and from 1000 to 5000 pounds of mercury annually (chiefly from nonpoint sources, including atmospheric fallout). Some 50,000 pounds of lead, mainly from point sources, find their way annually into Lake Michigan (Foran and Adler 1993, 34).

In an effort to eliminate such contamination, the International Great Lakes Water Quality Agreement identified 43 Areas of Concern, persistently polluted trouble spots, in the Great Lakes Basin. States are implementing remedial action plans to remove contaminated sediments in rivers and harbors and to restore "beneficial uses" that have been impaired. These include degraded fish and wildlife populations, loss of habitat, loss of sport and commercial fisheries, beach closings, and restrictions on drinking water. Participants in these remedial efforts strongly support pollution prevention plans to limit future contamination. The tasks are massive enough that they will take decades even with strong public support and adequate levels of funding.

One positive effect of these kinds of remedial actions plans is that the very scope of the problems and potential costs have provided much-needed incentives to adopt new approaches to environmental decision making. These include comprehensive or ecosystem management of affected areas. Such management addresses not only defined beneficial uses (which have an anthropocentric or human-centered focus) but also the integrity or health of natural ecosystems.

The Clinton administration has tried to make ecosystem management of this kind something of a trademark of its environmental policy, particularly for water quality and natural resources protection. For example, a massive and long-term plan to restore Southern Florida's Everglades will involve reshaping much of the region's 11,000 square miles to reestablish the natural flow of water through the area. It will involve extensive changes in farming areas, including development of 40,000 acres of filtration marshes to keep agricultural pollutants from the sugar industry from reaching the Everglades. Interior Secretary Bruce Babbitt has called this ecological restoration effort a crucial test of the administration's plan to manage whole ecosystems. Some 56 species in the region are listed as either endangered or threatened under the Endangered Species Act and all would be affected by the single, integrated plan.

Drinking Water

Drinking water quality across the nation is also a continuing problem. There is a widely acknowledged laxity of enforcement of regulations by the EPA and the states. The states complain that they lack sufficient funds and technical staffs to monitor water supplies as required for the more than 100 contaminants under Safe Drinking Water Act rules effective in 1995. Drinking water may contain a variety of dangerous compounds, including disease-causing microorganisms, lead, and chloroform (ironically, from the chlorine used to disinfect water supplies). In part because of lax enforcement, violation of federal health standards is common. An example of the effects could be seen in Milwaukee, Wisconsin, in 1993. A waterborne parasite, cryptosporidium, entered the city water supply and created an epidemic of intestinal disease, affecting more than 400,000 people and contributing to the deaths of more than 100 of them. According to the local press, thousands of residents had their confidence in the safety of local water so shaken that a year later they were still boiling their tap water or buying bottled water. The outbreak cost an estimated $54 million in health care expenses and lost productivity, and generated more than 1400 lawsuits against the city.

Several studies indicate that such problems with city (and private) water supplies are not as rare as commonly thought. The Natural Resources Defense Council reported in 1993 that it found violations by 43 percent of municipal water systems serving 120 million people (Terry 1993). The U.S. General Accounting Office (GAO) came to somewhat similar conclusions in a review of data for 1991, although it noted that 90 percent of systems in violation of drinking water standards served small communities and thus faced particularly difficult fiscal and technical problems in complying with all the standards (U.S. General Accounting Office 1994a). Millions of individuals not connected to municipal water systems rely on deep private wells that may pose comparable health risks, especially in agricultural areas where pesticides may contaminate drinking water supplies. Pesticide contamination is a major reason why the EPA has pressed Congress to revise federal law to allow imposition of more stringent limits on pesticide levels in water.

Another sign of the risks associated with use of well water came in 1994, when the EPA urged households with private wells to test for lead contamination after studies indicated that most submersible water pumps have brass or bronze parts that include lead. The lead could be released in high concentrations, particularly when the pumps are new. Consuming even very small amounts of lead can cause irreversible brain damage, especially in fetuses and young children. Lead poisoning is widely regarded as the most serious environmental problem facing the nation's children. For these and other reasons, in its comparative ranking of environmental risks, the EPA rated drinking water pollution as one of the top risks to human health. Many of the deficiencies noted here were addressed during congressional consideration of the Clean Water Act and the Safe Drinking Water Act in 1994. Both acts were on the congressional agenda again in 1995.

TOXIC CHEMICALS AND HAZARDOUS WASTES

Modern industrial societies like the United States are heavily dependent on the use of chemicals that pose risks to public health and the environment. They continue to produce vast quantities of them, although with some evidence of declining rates of production. The scope of the problem is captured in part in the quantities of chemicals produced and emitted into the environment each year, the amount of hazardous wastes produced as byproducts of industrial production and other processes, and the number of contaminated sites in the nation in need of cleanup or restoration.

Toxic Chemicals and Health Effects

The nation uses some 70,000 different chemical compounds in commercial quantities, and industry develops between 500 and 1000 new chemicals each year. U.S. production of synthetic chemicals burgeoned after World War II, increasing by a factor of 15 between 1945 and 1985. Similarly, agricultural demand for chemical pesticides such as DDT soared in the same period due to their low cost, persistence in the soil, and toxicity to a broad spectrum of insects. The EPA reports that use of pesticides nearly tripled between 1965 and 1985, with over six pounds applied per hectare (a hectare is 2.47 acres) in the United States by 1985 (Postel 1988). The EPA and the National Agricultural Chemicals Association indicated in 1993 that between 800 million and 1 billion pounds of some 600 pesticide ingredients are used each year in the United States (Schneider 1993b).

The overwhelming majority of these chemicals, about 90 percent, are considered safe, although most have never been tested for toxicity. Toxic chemicals are usually defined as a subset of hazardous substances that produce adverse effects in living organisms. To measure such toxicity in humans, we usually look to epidemiological data on the effects of human exposures (which may occur through pesticide residues on food, contaminated water supplies, or polluted air). Such data, however, are often incomplete or inclusive, making health effects difficult to establish. Only cancer has been studied extensively, and it may be too soon to detect effects from exposures over the past several decades. Researchers have estimated that from 2 to 8 percent of avoidable cancer deaths (i.e., those attributable to life style or environmental factors that can be modified) can be associated with occupation, 1 to 5 percent to pollution, and less than 1 to 2 percent to industrial products (Dower 1990). As Dower notes, even these low percentages, if correct, would translate into thousands of chemically related cancer deaths annually in the U.S. population.

Other less well-documented chronic health effects add to the problem. For example, a draft EPA report on dioxin in 1994 indicated that cancer may not be the most serious health hazard at common levels of exposure. Rather, concern focuses on the effects of dioxin on fetal development and the immune system, which may occur at very low levels of exposure, levels too low to lead to cancer (Schneider 1994c). Other health effects associated with toxic chemicals may include birth defects, neurotoxic disorders, respiratory and sensory irritation, dermatitis, and chronic organ toxicity such as liver disease. Scientists are giving more attention to possible synergistic effects, even at low levels of exposure, of diverse chemicals. These include pesticides and herbicides, heavy metals such as lead and mercury, and the ubiquitous PCBs and other chlorinated organic chemicals (Shapiro 1990).

There is ample scientific data to support reducing the use of toxic chemicals to mitigate such effects. Congress and the states have recognized the importance of such health risks in enacting policies directed at controlling undue exposure to harmful chemicals. Those policies now give us regular accounting of the nation's production and use of toxic chemicals if not their impact on human health and the environment.

Under the 1986 Emergency Planning and Community Right to Know Act, Title III of the Superfund law, manufacturers report annually to the EPA and to the states in which they have facilities the quantities of over 300 different toxic chemicals they release to the air, water, and land. The EPA records that data in its Toxics Release Inventory (TRI). In its report covering the year 1992, the EPA said that industry released some 3.2 billion pounds of toxic chemicals into the environment, or about 12.5 pounds per person.

Recent TRI reports indicate that about 60 percent, or nearly 2 billion pounds, went into the air and 7 percent into waterways. Industry injected about 20 percent into underground wells and deposited about 8 percent on land. The greatest contributions came from chemical manufacturing (accounting for about one-half of the total), and from production of primary metals, petroleum and coal products, paper, rubber, plastics, and transportation equipment. The top five most polluted states by this measure are Louisiana, Texas, Tennessee, Ohio, and Indiana. Louisiana and Texas, home to large chemical and oil refinery industries, accounted for fully one-quarter of the entire nation's releases of toxic chemicals. The release of such data had led many companies, such as Dow Chemical and Monsanto, to promise greatly reduced emissions of toxic chemicals.

In a related move in mid-1994, the EPA was preparing a rule to require any company in the nation making or using certain toxic chemicals to disclose the quantity it keeps at a facility and the health risks in the event of an accidental release. Such "worst-case" disclosures are expected to increase pressure on companies to reduce their inventories of toxic chemicals stored on site (Holusha 1994).

The EPA's latest report showed considerable progress since the first TRI data were collected in 1988, about a 37 percent reduction overall. Interpretations of progress are somewhat clouded, however, due to changing definitions of toxic chemicals, certain exclusions from the TRI database (including toxic wastes from mining, utilities, oil exploration, and wastes from small generators), and noncompliance by an estimated one-third of companies required to report. Environmental groups have complained in recent years that the TRI data understate the problem of toxic chemicals in the nation, and EPA administrator Carol Browner has indicated she plans to double the list of toxic chemicals subject to future TRI reports.

Handling Hazardous Wastes

The quantities of hazardous waste generated each year in the United States are also prodigious. Hazardous wastes are a subset of solid and liquid wastes disposed of on land that may pose a threat to human health or the environment with improper handling, storage, or disposal. Current federal law does not include all wastes with such risks. For example, the law excludes household wastes, agricultural wastes, mining wastes, and cement kiln-dust.

As is often the case, we have limited and conflicting data about the scope of hazardous waste problems, which are inherently less visible than air and water pollution and not always subject to current reporting. In the mid-1990s, the nation produced about 275 million metric tons of hazardous wastes annually, more than one ton per person. This was over 500 times as much as produced at the end of World War II, when the entire nation generated only 500,000 metric tons a year. These wastes come from an estimated 650,000 generators, the vast majority of them small facilities; about 2 percent of generators produce over 95 percent of the wastes. Most companies store these wastes at the point of generation, with about 5 percent sent elsewhere for disposal or some form of treatment (Dower 1990; Halley 1994).[4] Increasingly, however, industry is likely to recycle and otherwise reduce the production of hazardous wastes through pollution prevention initiatives.

The chief concern for hazardous wastes is leakage from corroded containers or unlined or leaking landfills, ponds, and lagoons, which can contaminate groundwater. This is an especially troubling problem because historically the wastes have been disposed of carelessly. Since 1950, over 6 billion tons of hazardous waste have been disposed of on the land, usually with no treatment and with little regard for the environmental consequences. As a result of such practices, among others, an estimated 30 percent of the nation's groundwater is contaminated to some degree, putting some 50 million Americans, chiefly in rural areas, at risk (Postel 1988). Similar concerns surround the special category of underground storage tanks (USTs) that contain petroleum or other hazardous chemicals. Congress added USTs to the Resource Conservation and Recovery Act (RCRA), the main federal hazardous waste law, in 1984. Recent estimates put the number of USTs in the United States at 1.7 to 2.7 million (Russell, Colglazier, and Tonn 1992; Cohen and Kamieniecki 1991). The EPA believes that 15 to 20 percent of tanks covered by the law are leaking or are expected to leak.

The chemical soup found in hazardous waste sites typically contains such dangerous compounds as trichloroethylene, lead, toluene, benzene, PCBs, and chloroform. It is impossible to generalize about health risks at each site due to variations in waste types and exposure. And as Roger Dower (1990) observed, while the potential health risks of exposure may be substantial, "little is known about the *actual* risks to the public from past and current disposal practices" (159). In 1986 Congress required the EPA to assess the risks to human health posed by each of the Superfund National Priority List (NPL) sites. The EPA and other federal agencies continue to study the risks, and credible data are likely to be available soon. The absence of such data, however, has fueled debate over the benefits that would accrue from the most inclusive and demanding cleanup policies.

Progress in cleaning up hazardous waste sites has been slow. The most frequently cited example is the federal Superfund program, which Congress created in 1980 following the highly publicized Love Canal chemical waste scandal in Niagara, New York. The Superfund legislation required the EPA to identify and clean up the worst of the nation's abandoned hazardous waste sites. Yet despite public and private spending of more than $13 billion through 1992, at only 149 of the 1275 sites on the NPL had all construction work related to cleanup been completed; only 40 had been fully cleaned up (U.S. Congress 1994). However, remediation is actively underway at hundreds of additional sites under the Superfund program, with about 65 to 70 being finished each year. Advances in scientific research and technological developments such as bioremediation, as well as statutory and administrative changes affecting site cleanups, are more impressive.

Even with such qualifications, it is clear that these limited achievements in dealing with the most troubling of the nation's hazardous waste sites come slowly and at a high price. In 1994 the Congressional Budget Office (CBO) estimated that the nation could spend about $230 billion through the year 2070 to clean up a total of 4500 nonfederal sites it expects to be placed on the NPL, or about $3 billion to $4 billion per year over the next 75 years; costs could be as high as $9 billion a year within the next decade (U.S. Congress 1994). In 1994 Congress worked extensively on a renewal of the Superfund program, well aware of the criticisms it had drawn (Camia 1994b). Although conflicts prevented approval of carefully negotiated legislation (see Chapter 4), revision of Superfund continued in the 104th Congress.

Contaminated Federal Facilities

One of the most prodigious tasks facing the nation is the cleanup of federal government facilities such as military bases and former nuclear weapons production plants. Those sites, while fewer in number than Superfund sites, are generally larger and present a more complex cleanup challenge, in part because of the mix of chemical and radioactive wastes (U.S. Office of Technology Assessment 1991a). Russell, Colglazier, and Tonn (1992) estimate that over the next 30 years, remediation activities by the Department of Defense (DOD) and the Department of Energy (DOE) could run between $110 billion and $430 billion, depending on the stringency of cleanup standards. Some 11,000 DOD sites may be in need of cleanup and over 4000 that are managed by DOE. The cost of cleaning up DOE's 17 principal weapons plants and laboratories alone will likely be more than $200 billion over the next several decades. At a spending level of some $10 billion per year, federal facilities cleanup already dwarfs EPA's operating budget (about $2.5 billion a year) and greatly exceeds annual Superfund cleanup costs. In late 1993 the Clinton administration created a Federal Facilities Policy Group to coordinate a strategy for assessing risks and for planning and managing the long-term cleanup effort.

Radioactive Wastes

The disposal of high-level radioactive wastes from commercial nuclear power plants represents a comparable problem in the handling of dangerous waste products. It has proved to be equally difficult to resolve. By 1994 the United States had accumulated about 27,000 metric tons of high-level wastes, chiefly spent fuel from power plants. It was adding 2000 metric tons each year from the continued operations of 109 commercial reactors. For the world as a whole, the annual production of high-level waste is about 8000 tons a year. The U.S. Department of Energy has estimated that the nation will have about 59,000 metric tons of such waste by the year 2010, and 80,000 metric tons by the year 2025, assuming that no additional nuclear power plants are built. The spent fuel rods remain highly dangerous for thousands of years. Most of the radioactivity decays quickly, but enough remains that the EPA standards for disposal of this kind of waste specify 10,000 years of isolation from the biosphere to protect public health and the environment.

These high-level wastes have been stored since the 1950s in water-filled basins at reactor sites around the nation. Currently the wastes are at 70 commercial nuclear plants located in 33 states. As space runs out, the future of the nuclear industry depends on finding more permanent locations for the waste. The industry is eager to have such a solution. In the Nuclear Waste Policy Act of 1982, Congress mandated construction of permanent geologic disposal sites. After a 1987 amendment to the act, attention focused exclusively on a site at Yucca Mountain, Nevada. However, technical uncertainties over how to safely isolate the waste from the biosphere for 10,000 years, mismanagement of the program by the DOE, and especially political opposition in Nevada and elsewhere to the siting of such a facility have stalled the policy. Public fear of radioactive waste and distrust of DOE are great enough that assurances of safety are widely disbelieved (Dunlap, Kraft, and Rosa 1993). Proposals for interim storage of the waste in large steel and concrete casks at reactor sites have run into similar opposition by communities, states, and environmentalists, although such casks already are being used at six U.S. plants. Failure to resolve the conflicts may well end the use of nuclear power, forcing greater reliance in the short term on fossil fuel plants even if in the longer term alternative sources of energy are promising.

SOLID WASTE AND CONSUMER WASTE

An early concern of the environmental movement dealt with the byproducts of the consumer society that clogged municipal landfills. The problem continues. EPA estimated that in 1990, Americans produced about 196 million tons of municipal solid waste, more than double the 88 million tons produced in 1960, and considerably higher than the 151 million tons in 1980. For 1990, this represented 4.2 pounds per person per day, or over 1500 pounds a year per capita (Council on Environmental Quality 1993). This is roughly twice the waste per capita generated in Western European nations or Japan. These amounts are growing at an estimated 1 percent a year, slightly outpacing U.S. population growth. About 40 percent of that waste is paper and paperboard, 25 percent is yard waste and food, 8.5 percent metals, 8 percent plastics, and 7 percent glass (World Resources Institute 1992). These household, institutional, and commercial wastes are only the tip of the solid waste iceberg. The vast majority of solid waste comes from industrial processes, including agriculture and mining, raising the total to some 13 billion tons per year.

Compounding the problem of disposing of such huge quantities of wastes is the sharply declining number of municipal landfills open to receive them. They fell from 20,000 in 1978 to only 6000 by 1986 (Council on Environmental Quality 1993). The EPA estimates that by the year 2009, perhaps 80 percent of the remaining landfills will close, either because they have reached capacity or they cannot meet environmental standards. Opening new landfills has proved difficult due to community opposition and stringent new requirements for construction and operation (e.g., to control air pollution). As a result, disposal costs have escalated sharply in some parts of the nation, particularly in the Northeast, where they reach $100 per ton or five times the cost in other areas. On a positive note, these developments have created opportunities for source reduction, the best alternative, and reuse, recycling, and composting of solid waste, or its use for generation of energy.

The EPA reported that in 1990 that the nation recovered through recycling or composting about 33 million tons of municipal solid waste or about 17 percent of the amount generated. Almost two-thirds of the waste saved was paper and paper-board. In 1994, cities incinerated about one-fifth of municipal solid waste in the na-tion's 128 waste-to-energy plants. Yet the outlook for such incineration is uncertain because of the toxic pollutants it produces, including dioxin. In addition, legal chal-lenges, lower energy prices, and the availability of landfills may lead to alternative methods for the disposal of solid waste.

One special concern is the generation of incinerator ash that may contain haz-ardous waste and the cost of its disposal. In *Chicago v. Environmental Defense Fund,* a case brought by the Environmental Defense Fund against the city of Chicago (in which the EPA sided with the city), the U.S. Supreme Court ruled in 1994 that any toxic residue from the burning of household and industrial waste in in-cinerators must be tested for hazardous waste characteristics. If found to be haz-ardous, the residue cannot be put in ordinary landfills.

The effect of this ruling may be to sharply increase the cost of incinerator opera-tions, depending on whether cities find it feasible to separate the trash responsible for the toxic residue, largely discarded batteries, paint cans, and electronic equip-ment. Environmentalists argue that the court's decision creates strong incentives for municipalities to increase their pollution prevention programs to reduce what would otherwise be prohibitively high disposal costs. Officials of the National League of Cities have asked Congress to revise federal law to set less strict standards. Recent EPA studies have indicated that proper disposal of hazardous wastes in certified fa-cilities can be over three times as costly as using ordinary landfills. Estimates by some cities suggest the costs may be as much as ten times higher, or anywhere from $200 to $500 per ton compared to $30 to $50 per ton for regular landfills (Schneider 1994a).

The amount saved from landfills doubled between 1985 and 1990, and compara-ble savings should grow as more local governments launch recycling programs. Some 30 states have comprehensive recycling laws, and some innovative programs, such as the one adopted by the city of Seattle, have become models of effective curbside recycling, with economic incentives to ensure widespread citizen coopera-tion. Other actions, including enactment of the federal Pollution Prevention Act of 1990, the creation of new markets for recycled goods, and tightening restrictions on disposal of hazardous waste, should reduce industrial waste quantities as well.

ENERGY USE AND CLIMATE CHANGE

Energy use has a major effect on most of the environmental problems discussed above, especially air pollution, acid precipitation, and production of greenhouse gases. It also affects both the health of the economy and national security. Despite these consequences, the United States has never found it easy to address energy problems and policy proposals.

The Scope of Energy Problems

Energy problems may be defined in part by the total amount of energy used, the effi-ciency of use, the mix of energy sources relied on, and the reserves of nonrenewable sources (e.g., oil) that remain available. Among the most important considerations are the environmental costs associated with the life cycle of energy use: extraction, transport, use, and disposal. When oil is transported, spills may occur, sometimes spectacular ones like the 10 million gallons of crude oil that leaked from the Exxon *Valdez* supertanker in Prince William Sound off Alaska in 1989. When the fuel rods that power nuclear plants are "spent," they must be disposed of as highly radioactive waste. The effect that most concerns students of energy policy is the possibility of global climate change due to the buildup of greenhouse gases, discussed briefly in Chapter 1. The most consequential greenhouse gas is carbon dioxide produced in the burning of fossil fuels. The United States releases more greenhouse gases (5.2 billion metric tons in 1990) than any other nation, the vast majority of those emissions coming from fossil fuels that power automobiles, make electricity, run industrial processes, and heat homes.

An overview of selected data on energy use conveys a simple message. It would be hard to overstate the importance of the world's choice of energy paths for the future. At the same time, energy use is so closely tied to vital industrial processes and highly valued public conveniences such as air conditioning and automobile use that it is also easy to understand the difficulty nations have in trying to alter that path in the short term.

For the latest year for which full data are available (1991), the United States derived over 88 percent of its energy from fossil fuels (coal, oil, and natural gas), and about 12 percent from other sources, including hydroelectric (3.4 percent) and nuclear (8.1 percent) (U.S. Department of Energy 1993).[5] Thus fossil fuels and nuclear power account for over 96 percent of total energy production in the United States, and use of each leads to substantial adverse environmental impacts. Sources such as geothermal, solar thermal and photovoltaic generation, and wind power are still a distinctly minor part of the energy picture, though with prospects improving greatly in recent years (Alliance to Save Energy 1991; U.S. General Accounting Office 1993a).

In 1992 the United States imported 46 percent of the oil it consumed at a cost of $53 billion a year, with major consequences for national security and the international trade deficit. Hazel O'Leary, secretary of energy, has estimated that by 2010, U.S. dependency on imported oil could reach 60 percent, at a cost of $130 billion per year (Balzhiser and Bryson 1994). Continued low prices keep the nation dependent on imported oil. In 1990, the United States used 16.9 million barrels of oil per day. The nation burned about 43 percent of it in automobile engines, and another 20 percent in trucks and airplanes. In 1994, adjusted for inflation, the cost of oil imported from OPEC nations was about the same as it was in 1973 before the Arab oil embargo forced energy issues onto the national agenda. Similarly, adjusted for inflation, the cost of gasoline in the United States in 1994 was as inexpensive as it was before the nation discovered the energy issue, although in most other industrialized nations it was three to four times as costly. Cheap oil and gasoline discourage conservation and investments in efficiency, and make development of alternatives, including domestic sources of oil, more difficult. U.S. dependency on oil imports and the nation's enormous appetite for oil create periodic conflicts over access to domestic oil fields such as those underlying part of the Arctic National Wildlife Refuge

(ANWR), the development of which environmentalists fought bitterly and success-fully in opposing President George Bush's National Energy Strategy in Congress in 1991.

Projection of future energy use is difficult because so much depends on developments in energy efficiency, new technology, market prices, government policies, and changing consumer behavior. For example, the nuclear power industry would desperately like to see a major expansion using new reactor designs that promise cheaper and safer nuclear energy. Yet public opposition and utility resistance remain strong, costs are uncertain, and the nuclear waste issue continues to hamper expansion of the industry.

Nevertheless, there are a number of contrasting long-term (20- to 40-year) scenarios for U.S. energy use. Most assume continued emphasis on fossil fuels, a major but stable role for nuclear power, and a slow transition to renewable energy sources as economic forces and technological developments permit. Over the next 20 years, the U.S. DOE projects increased demand for oil (largely imported as U.S. production is decreasing), a continued rise in coal use to keep pace with growing demand for electricity, and an increase in use of natural gas. Studies by environmental organizations indicate that if current policies and energy use trends continue until 2030, U.S. energy consumption will rise by 41 percent and carbon dioxide emissions will increase by 58 percent from 1988 levels.

Other energy scenarios, however, indicate the feasibility of significant reductions in energy use as well as in air pollution, and a shift to sustainable energy sources such as solar, geothermal, wind, and hydrogen power. Several environmental groups have estimated that the nation could lower fossil fuel use to no more than 42 percent of its energy mix by 2020, with as much as 50 percent coming from sustainable sources. Following these kinds of energy paths could cut U.S. energy consumption in half and reduce carbon dioxide emissions by 70 percent at a cost of $2.7 trillion. But the plan would also save $5 trillion in fuel and electricity costs over the next 40 years, producing a net savings for the economy of $2.3 trillion (Alliance to Save Energy 1991).

Fossil Fuels and the Threat of Climate Change

As noted earlier, the world's scientific community and numerous independent policy studies support the basic case for rapid reduction in use of fossil fuels, primarily because of their contribution to possible climate change (Intergovernmental Panel on Climate Change 1990; National Academy of Sciences 1991; Hempel 1993). To date, such proposals have suffered for lack of political support. Even very modest proposals for increasing the gasoline tax meet stout opposition in Congress. Such appraisals of political acceptability help explain President Bill Clinton's limited Climate Change Action Plan announced in October 1993.

Under the UN Convention on Climate Change approved at the 1992 Earth Summit, the United States and other developed countries that agreed to the treaty must cut greenhouse gas emissions to 1990 levels by the year 2000. Clinton favored a plan that relies on voluntary efforts and business-government partnerships rather than new taxes or regulations. Through some 50 cooperative projects such as encouraging utilities to use alternative energy sources, the administration anticipated a cumulative energy savings of $200 billion by 2010. Environmentalists argued that the Clinton plan was insufficient to reduce greenhouse gases to ecologically safe levels even if all of its actions were taken.

A 1994 report from the Intergovernmental Panel on Climate Change strengthens the case for taking more aggressive action in the short term. The panel found that even if the climate change convention targets for reduced emissions were applied globally and fully met, carbon dioxide levels would still continue to increase for two centuries. Just to keep atmospheric concentrations of carbon dioxide from rising to twice today's levels would require cutting emissions *below* 1990 levels (Stevens 1994b). At a UN-sponsored meeting in September 1994, delegates from the leading industrial nations, including the United States, conceded that they are unable to meet even the 1992 climate convention goals without additional measures (Washington Post 1994). Environmentalists continue to maintain that the single most important action to cut emissions further would be increased federal fuel efficiency standards.

Substantial gains in U.S. energy efficiency have been made over the past 20 years, and they continue to be made. In 1993 the nation used about 10 percent more energy than it did in 1973, but there were 30 percent more homes, 50 percent more vehicles, and a 50 percent increase in the gross domestic product (Council on Environmental Quality 1993). Even so, Americans use twice the energy per capita as Germany and Japan, and thus waste energy and absorb added fuel costs estimated to total $200 billion per year. Further gains can be expected as the Energy Policy Act of 1992 is implemented. The act mandated greater energy conservation and efficiency and increased support for research and development. The EPA's new Energy Star program has already encouraged the development of a new generation of computers with power-saver features. Several major environmental groups, including the Natural Resources Defense Council and the National Audubon Society, have demonstrated the potential of energy efficiency through extensive use of new energy-saving technologies in the remodeling of their headquarters buildings.

As these examples illustrate, more aggressive efforts than commonly taken are clearly possible. Southern California demonstrated recently that such energy paths can be politically acceptable as well. In the early 1990s, the state, and especially the Los Angeles metropolitan area, embarked on a remarkable experiment in integrating energy and environmental policy that promises to meet future energy demands in Southern California almost entirely through conservation, efficiency, and renewable energy sources and simultaneously to reduce pollution levels (Mazmanian 1992).

The outlook is much less favorable on a worldwide basis, where energy demand is likely to increase sharply under almost any scenario, driven by rapid population and economic growth in the developing nations. Unless nations develop alternatives, those demands are likely to be met largely with fossil fuels. The U.S. DOE reported in 1994 that developing nations already have become the world's leading producers of carbon dioxide. As of 1992, they produced 52 percent of global energy-related carbon dioxide emissions, about a 25 percent increase over levels produced in 1970. Even in the developing nations, however, alternatives to fossil fuels exist if nurtured by political leadership and public policy (Lenssen 1993; Holdren 1991).

BIOLOGICAL DIVERSITY AND HABITAT LOSS

Biological diversity, or biodiversity, refers to the variety and variability among living organisms and the ecological complexes in which they occur. The tangible things called biodiversity and the habitats that support them derive from the ecological and evolutionary processes that have shaped them over geologic time spans, and will continue to do so in the future.

We humans have long taken biodiversity for granted and have enjoyed the free services it has provided for us. Yet human activity, both intentional and inadvertent, has had a devastating effect on biodiversity over the past 10,000 years. People have always cleared land, and have over-hunted some species and caused extinction of others. The present is distinguished from the past primarily by the magnitude of destruction and rates of change. The principal causes of the new threats to biodiversity are habitat loss and modification (such as fragmentation), pollution and contamination, overexploitation of species and habitats, introduction of exotic and competitive species, and synergistic and cascading effects deriving from these activities. In the broadest terms, the goal of sustainable environmental management is the perpetuation, or restoration, of the full range of biological entities, and particularly those dominated by evolving processes of life, and the maintenance of their reproductive, regenerative, and adaptive capabilities (Barnthouse 1995).

Biodiversity Loss and Implications

Richard Tobin (1990) notes that between the years 1600 and 1900, human activities led to the extinction of about 75 species of birds and mammals, or about one species every four years. A comparable number was lost in the first half of the twentieth century. Biologists estimate that in the mid-1970s, anthropogenic, or human caused, extinctions rose to about 100 species per year. Edward O. Wilson, one of the nation's leading authorities on biodiversity, believes the extinction rate in the mid-1980s was accelerating rapidly and was at least 400 times the natural rate, perhaps as high as 10,000 times the rate that existed prior to the arrival of human beings (Wilson 1990, 54). Even the congressional Office of Technology Assessment reported in a major 1987 study that the loss of biological diversity was of "crisis proportions." That view does not seem to be widely shared by the American public and its elected representatives (Tobin 1990).

The impact of such species loss is particularly acute in tropical rain forests. Although there is no certainty about the number of species in existence, biologists have estimated that more than one-half of all species live in moist tropical forests. Such forests cover only 6 percent of land area but are rich biologically, especially with insects and flowering plants. The forests are especially vulnerable ecosystems, and they illustrate the larger threat to biodiversity of human interventions.

By some estimates, from the dawn of agriculture to the present, humans have eliminated about 20 percent of the world's forest cover. The figure is higher in tropical forests; some 55 percent of the original forest cover has been lost. Recent calculations put the loss at an estimated 1.8 percent of the total rain forests annually, or about 140,000 square kilometers per year. This is roughly the size of the state of Florida

(Ehrlich and Wilson 1991). In the United States, some 60,000 acres of ancient or old growth forest had been cut each year before recent logging restrictions (Wilson 1990).

The reasons for this loss of forests are not in much dispute. Among the major factors are the clearing and burning of rain forests to make room for rapidly growing and poor populations, conversion of forests for planting of cash crops and cattle pastureland, commercial logging, overharvesting of fuelwood, and dam construction. Many of these and similar activities have been encouraged by short-sighted government policies (Miller, Reid, and Barber 1991; Tobin 1994).

As the forest habitat is destroyed, species are lost. Erhlich and Wilson (1991) have estimated that the annual loss of rain forests translates into the permanent loss of about 4000 species each year. On the basis of such estimates, 25 percent of all species now in existence would be gone by the year 2015 if these rates continue. The world would then be losing 15,000 to 50,000 species per year, or 50 to 150 per day. Recent studies indicate that global warming is also likely to adversely affect rain forests, not so much through temperature increases as through disruptions in weather patterns.

There are many reasons to worry about such species loss. Forests provide us with opportunities for recreation and for aesthetic enjoyment. They also are a treasure trove of medicinal drugs, oils, waxes, natural insecticides, and cosmetics, and they could contain future sources of food (World Commission on Environment and Development 1987). Of the 80,000 edible plants, humans have used about 7000 for food, but we actively cultivate only about 200, and rely heavily on only about 20— such as wheat, rye, corn, soybeans, millet, and rice (Wilson 1990, 58). As important as those ecological and agricultural values are from an anthropocentric perspective, they are not as compelling as arguments advanced from an ecocentric or ecology-centered viewpoint.

Biodiversity, whether in tropical forests or elsewhere, is important because it provides irreplaceable ecological values, including the genetic heritage of millions of years of evolution. We risk damage to the functioning of ecosystems with species loss, and the permanent disappearance of diverse genetic codes that could prove invaluable for species adaptation in what may be a rapidly changing environment in the future. Ecologists continue to debate the precise relationship between biodiversity and ecosystem health or productivity—that is, how loss of species affects ecosystem functions (Baskin 1993). Yet there is no disagreement that the functions put at risk with loss of biodiversity are critical: the cycling of nutrients, control of climate, regulation of fresh water supplies, and maintenance of soil fertility and atmospheric quality. The issue is not, as the press reports it, whether a single species, such as the northern spotted owl, is lost. Rather it is that we risk the destruction of critical habitats and ecosystems on which life depends, including human life.

Policy Actions and Effects

Protection of biodiversity was one of the most controversial issues at the 1992 Earth Summit, which also produced an accord on protection of the world's forests. President Bush refused to sign the proposed biodiversity treaty at the meeting, and he expressed concern about rights to commercial exploitation of biological resources and other issues. However, in April 1993, following discussion with business and environmental groups, President Bill Clinton signed the treaty. The Convention on Biological Diversity took effect in late 1993, with the United States among the 167 nations supporting it. By the end of 1994, the U.S. Senate had yet to ratify the agreement, but its eventual approval would not seem to be in much doubt.

Solutions to biodiversity problems are not scarce even if the political will to adopt and enforce them is. For developing nations, policy makers and environmental groups propose swapping international debt for preservation of natural areas, land reform to promote greater equity and thus sustainable use of land, integrated land use planning, an end to governmental subsidies of deforestation, and measures to slow or halt population growth.

In the United States, protection of biodiversity is tied closely to enforcement of the controversial 1973 Endangered Species Act. Although the act has a broad mandate for ecosystem conservation, emphasis to date has been on protection of individual species. Even here, the act has achieved only modest success after more than 20 years. Some 700 species have been listed as threatened or endangered, and in recent years the federal government has added about 50 new species each year to the list. The Fish and Wildlife Service (FWS), which implements the act for terrestrial and some aquatic species, designated scores of critical habitats and put many recovery plans into effect. Unfortunately, only a few endangered species have recovered. The slow pace of progress may be attributed in part to small budgets and insufficient staff for the FWS. Political opposition also has been important, although most species preservation efforts have moved forward without the high-visibility opposition that surrounds protection of old growth forests and the spotted owl (Tobin 1990). Over 3000 species are candidates for inclusion on the list, but without additional resources the FWS cannot quickly assess their status. In late 1992, in a legal settlement with environmental groups, the FWS agreed to propose for listing within four years the 400 top-ranked candidate species. The agency also promised to expedite consideration of 900 other species where definitive scientific data are lacking (Council on Environmental Quality 1993). Increasingly, emphasis is expected to be given to ecosystem, rather than single species, preservation.

In the mid-1900s, several large programs were actively building a knowledge base for further protection of U.S. biodiversity. These include the Department of Interior's National Biological Service, the EPA's Environmental Monitoring and Assessment Program (EMAP), the FWS's gap analysis program, and the Nature Conservancy's Biological and Conservation Data System. The last is a private biogeographic database that permits assessment of species diversity by regions to establish protection priorities. Eventually such data will be integrated in Geographical Information System-based profiles of national biodiversity resources that can serve as the basis for identifying critical areas in need of greater protection.

HUMAN POPULATION GROWTH

A long-standing debate posits that either population growth or the use of technology causes environmental degradation. Thus, some have pressed for curtailing the world's high population growth rate through family planning and social and economic development. Others have emphasized the need to regulate the use of polluting technologies. Mainstream environmental organizations such as the National Audubon Society, the National Wildlife Federation, and the Sierra Club that had long ignored population growth rediscovered it in the late 1980s. Many actively promoted the urgency of action on this issue in conjunction with the UN Conference on Population and Development held in September 1994 in Cairo.

Population and Sustainable Development

Dealing with both population growth rates and the use of technology is essential to achieving sustainable development; each contributes to degradation of the environment. A high rate of population growth makes sustainability almost impossible to achieve given the necessity of providing adequate food, land, water, energy, and other essentials of life. The natural carrying capacity of the land (which is determined in part by the technology used and levels of consumption and waste generation) may be exceeded in the process (Mazur 1994). The cost in human lives is also great. An estimated 40 million people die each year from hunger or hunger-related diseases, primarily in developing countries. Almost 40,000 children under the age of five die each day in poor nations from treatable diseases such as malaria, diarrhea, measles, tetanus, and acute respiratory infections that rarely kill Americans (Tobin 1994). In addition, high population growth rates may affect political stability within nations and conflict among them if battles erupt over scarce natural resources such as water and arable land. Sustained economic growth, national security, social peace, and human justice depend on limiting and eventually halting human population growth, improving scientific knowledge and technological systems, and reallocating critical resources such as land and water to more efficient and equitable uses.

Birth rates have declined since the 1960s, but they remain well above the replacement level (slightly more than two births per woman of child-bearing age) that eventually produces a stable or nongrowing population. Projections of future population, both globally and in the United States, provide little basis for complacency, either for economic development in poor nations or for protection of critical environmental resources worldwide. The impact on habitats and biodiversity, air and water quality, energy and water use, and other aspects of environmental quality will likely be enormous, but also geographically varied. Some nations and regions of the world will suffer far more than others from food shortages, poor health, environmental degradation, and economic dislocations, including widespread unemployment. In sub-Saharan Africa, for example, one-third of the population already lives in abject poverty, unable to meet basic needs. With a growth rate of 3.0 percent a year, half the population may live in such circumstances by the year 2000.

Considerable disagreement exists over the possibility of feeding and otherwise providing for the needs of Earth's burgeoning human population, and over what the ultimate carrying capacity of the planet is. There is less dispute over the imperative of intensive agricultural research and technology development to try to meet the surging demand for food, and over a general strategy of emphasizing early action and prevention rather than adaptation to scarcity after the fact.

We can hardly place all the blame on the poor nations, where 95 percent of the future growth in human numbers will occur. The industrialized nations consume a vastly greater proportion of the world's resources and have a much higher per capita impact on the environment. The richest one-quarter of the world's nations consume 60 percent of the food and use about 70 percent of the metals and energy and 85 percent of the wood. They also generate 90 percent of the industrial and hazardous wastes, and use 80 percent of ozone-depleting CFCs. The United States, with 5 percent of the world's population consumes 25 percent of its commercial energy and produces 18 percent of global greenhouse gases. Those numbers suggest what the future might hold if high global population growth rates are combined with intensive economic development based on the technologies currently in use in the developed nations. They also indicate a moral imperative for developed nations to lower energy and materials consumption.

Growth Rates and Projected Population Increase

In mid-1994, the world's human population was growing at 1.6 percent annually, adding to its base of 5.6 billion about 88 million people every year. This is a net increase in the world's population (not the number born) of nearly 240,000 people per day. The population of the poorest nations is expected nearly to double over the next 30 years (Population Reference Bureau 1994). Although fertility rates continue to decline slowly, the United Nations' medium, or most likely, projection for the year 2000 is 6.3 billion people and for 2050, 9.8 billion. Its high series projection puts the total at 11.9 billion in 2050 and the low series at 7.9 billion (United Nations 1994).[6]

All such projections depend critically on assumptions about economic and social development, the availability of family planning programs, and the extent of contraceptive use. Social development—such as improved education, a higher status for women, better health care and nutrition, adoption of old age security programs, and economic reform—is as important as provision of family planning services. According to a recent World Bank study, family planning programs accounted for about 40 percent of the decline in fertility in developing nations; about 54 percent of the decline was attributed to socioeconomic development (Keyfitz 1990). Yet even family planning is not reaching all who want it. Up to 350 million couples worldwide want family planning services and cannot get them.

The United Nations has recognized both the continuing need for family planning and the imperative of social and economic development. It incorporated both kinds of programs into its recommendation for the Cairo conference. The undersecretary of state for global affairs, Tim Wirth, coordinated U.S. preparations for the conference and endorsed the same position. He called it the "new consensus" and the "cornerstone of the Clinton administration's population policy" (Mazur 1994, xv). In some nations, the combination of such efforts has produced striking declines in fertility levels. These include Cuba, Sri Lanka, South Korea, Thailand, Tunisia, Singapore, Mexico, Indonesia, and China. Other nations have made much less progress toward lower growth rates. An added factor for developing nations is rapid urbanization of the population, leading to overcrowded and severely polluted "megacities."

In contrast to the world average of 1.6 percent, the population of developed nations is growing by an average of only 0.3 percent per year, creating what some have called a demographically divided world. Western Europe has an average growth rate of only 0.1 percent. The U.S. rate, however, is about three times the average for developed nations, or about 1 percent a year counting immigration. The U.S. population (approximately 261 million in mid-1994) increases by some 2.5 million people per year. Growth is likely to continue for at least another six decades, even if the fertility rate remains below the replacement level, due to a high level of immigration and recent increases in fertility. The U.S. Census Bureau estimates that the nation will reach 300 million by the year 2010 and 392 million by 2050 (Population Reference Bureau 1994; Kraft 1994b).

The impact in selected areas of the nation such as Florida, Arizona, and Southern California, or in rapidly growing cities elsewhere, is often dramatic. In the early 1990s, Florida was gaining over 6000 residents every week. The city of Las Vegas, which has experienced explosive growth in recent years, is now home to 1 million people, and continues to add 1000 residents each week—putting enormous demands on area water supplies. Cities and states will have to plan carefully to minimize adverse impacts on land, water supplies and water quality, air quality, critical habitats, urban infrastructures, and the overall quality of life as population grows and congestion increases. In this context, it is notable that the United States has no direct policies to encourage distribution of the population, relying instead on private actions and the marketplace. The nation also has no explicit population policy, a goal that Earth Summit participants endorsed for all nations. The chapter on "demographic dynamics and sustainability" in the conference's Agenda 21 action plan recommended, among other things, incorporation of population issues into environmental planning, more research on environmental carrying capacity, and adoption of national population policies to help establish paths toward sustainable development.

CONCLUSION

This selected overview of U.S. and global environmental problems provides at least some indication of the scope and severity of current threats to public health and ecosystem integrity. Debate continues on the extent of environmental and public health risks and about the kind of data that should be used to answer those questions. Federal agencies and international environmental organizations recognize the inadequacy of present monitoring of environmental trends, and the need to improve environmental data collection, its integration, and its assessment.

The proposed Department of the Environment would likely include a Center for Environmental Statistics that would issue regular reports on environmental conditions and trends, either as a supplement to or a replacement for the Council on Environmental Quality (CEQ). The CEQ was given those duties by the National Environmental Policy Act of 1969, though it has never had sufficient staff or resources to fully meet those expectations. The center could serve as a clearinghouse on environmental data, drawing from the EPA's EMAP program, the U.S. Geological Survey's

National Ambient Water Quality Assessment, and similar efforts by other federal agencies, the states, and private organizations.

In 1992 the EPA identified 83 different environmental data programs in 25 separate federal agencies—hence the need to integrate the various streams of data and to improve the nation's ability to assess interrelationships among different environmental stressors and overall environmental quality, and to relate the findings to public policy actions. The center also could help to develop the tools needed for long-range forecasting of environmental trends and could issue regular reports on progress toward long-term goals (National Commission on the Environment 1993). At a time when the rate of environmental change threatens to outstrip our capacity to assess and respond to it, there is no substitute for improved foresight. Without a substantial increase in support for environmental research, however, inadequate scientific knowledge and assessment of it will continue to hobble environmental policy making. Total research and development (R&D) spending on the environment in recent years has been less than $5 billion annually (of which less than 10 percent goes to the EPA) out of a total federal R&D budget of nearly $75 billion.

Partly because of the paucity of available data, there continue to be disputes over the extent of progress being made in dealing with air and water quality, toxic chemicals and hazardous wastes, and most of the other problems summarized above. Disagreement is particularly intense over how much more should be done, with what policy instruments, and at what cost. We will return to these questions in Chapter 6 on evaluating environmental policy. Some conclusions are clear enough, however. Environmental policy efforts in the future will have to focus far more than they have to date on specific and measurable indicators of progress. The nation (and world) will also have to assess and compare the diversity of environmental and public health risks in order to set priorities for action. And we will have to do a better job of selecting the policy approaches (for example, regulation, market incentives, or public education) that promise to work best.

There is another message for all students of environmental policy. We need to improve our individual and collective capacities to review and judge the scientific facts and the political assertions tied to them and to learn better how to bring environmental knowledge to bear on policy choices. Citizens need to be more alert to the sometimes uncertain scope of environmental problems and their consequences and to be prepared to play an active role in deciding what to do about them, from their local communities to the state, federal, and international level. As we will see in the next chapter, public opinion on the environment has been a powerful driver of environmental policy in the twentieth century. But as environmental issues become more complex and the policy choices more contentious, generalized public support will be a weak deterrent against the determined efforts of organized interests to block specific policy actions that affect them adversely. If citizens want to shape the direction of environmental policy in the 1990s and in the twenty-first century, they will have to become far more knowledgeable and be ready to take advantage of the many opportunities available to influence governmental and corporate decisions.

ENDNOTES

1. In addition to covering the medical consequences of urban and transboundary air pollution, Chivian et al. (1993) review evidence of human health consequences of the decline in drinking water quality, food contamination, occupational exposures to toxic agents, radiation, ozone depletion, climate change, population growth, and loss of biological diversity.

2. Control of gasoline-powered lawn mowers may strike some as unnecessary, but they are a significant and previously unregulated contributor to air pollution. The California Air Resources Board reports that a dirty power mower can emit as much pollution in 30 minutes of operation as a modern automobile does in 172 miles of driving. The federal EPA estimates that lawn mowers alone account for about 5 percent of air pollution; all nonroad engines together produce between 14 and 19 percent of air pollution. In 1994 the EPA announced new federal rules for gas mowers and about 65 other kinds of nonroad gasoline-powered equipment that are to take effect in 1996 as part of the implementation of the 1990 Clean Air Act, though they will apply only to newly manufactured equipment.

3. The EPA report is entitled *Respiratory Health Effects of Passive Smoking: Lung Cancer and Other Disorders.* The EPA's staff strongly defends its 530-page health risk assessment against tobacco industry charges of inadequate science. The report was four years in the making, involved an extensive review and evaluation of several hundred scientific studies on ETS. Two drafts of the report were reviewed externally and public comments were received on it. The SAB committee (consisting of 18 independent experts) held two public meetings on the report, and concurred in the EPA's methodology for consolidating existing health risk assessments. The committee members also unanimously endorsed the EPA's classification of ETS as a Group A carcinogen (Bayard and Jinot 1993).

4. Halley (1994) reports that the RCRA-regulated universe in 1994 consisted of 4700 hazardous waste treatment, storage, and disposal facilities (with some 81,000 individual waste management units) and 211,000 facilities that generate hazardous waste. Over the past decade, the number of generators covered by RCRA has increased ninefold, largely because of the reduced threshold for waste quantities for inclusion in this category.

5. The Energy Information Administration in the Department of Energy publishes a comprehensive *Annual Energy Review* that surveys the nation's production and consumption of all major energy sources.

6. A recent review of population forecasting methodologies, with several scenarios that differ from UN estimates, can be found in W. Lutz (1994). Lutz's "central scenario" projects a world population of 9.5 billion in 2030 and 12.6 billion in 2100.

CHAPTER THREE

U.S. Politics and the Evolution of Environmental Policy

Environmental problems in the 1990s occupy a prominent position on national and international political agendas even as debate continues over their severity, the level of governmental intervention needed, and the policy approaches that work best. The 1992 Earth Summit testifies to their status as does the extensive debate given to environmental policy questions in the 1992 presidential election. It was not always thus. Prior to the late 1960s, environmental issues were barely mentioned in the national media and were of little interest to most policy makers. The pattern continued even as evidence of environmental degradation mounted during the 1960s. The salience or importance of the environment rose rapidly in the late 1960s. Since then it has ebbed and flowed in response to shifts in the economy and political climate.

These fluctuations speak to an important aspect of environmental policy making. The mere existence of detrimental environmental conditions and dire warnings of future trends provide no guarantee that governments will respond. The problems must achieve a sufficient level of visibility and thus command political attention. This is what is meant by getting on the agenda. When governments do act, the particular form of public policy chosen may or may not be the best way to address a given problem, and policy priorities may not reflect the relative risks to which the public is exposed. Politics has a great deal to do with such choices and effects: the extent of public concern and involvement; the activities of environmental, business, and other interest groups; and the weight policy makers give to different arguments and proposals.

Designing appropriate responses to environmental problems requires an understanding of the political process, and especially how problems get defined and move onto political agendas and stay there or fall off, and how proposals are developed, debated, and transformed into environmental policy. In this chapter I turn first to the policy-making process and some general features of U.S. government that shape environmental politics. I then discuss the evolution of natural resource and environmental policy, with emphasis on important changes in the twentieth century, to provide at least some basis for judging how well the U.S. government has responded to environmental challenges.

THE POLICY-MAKING PROCESS AND THE ENVIRONMENT

The U.S. policy-making process may strike the uninitiated as inordinately complex if not completely mystifying. This is especially so when decisions focus on narrow and technical issues understood only by those intimately involved with the policy. Few people have the time or inclination to follow the intricacies of nuclear waste policy, such as scientific assessments of potential waste repository sites, or comparable aspects of clean air policy or drinking water standards. Yet the overall character of policy making and the stages through which most environmental policies move is straightforward and easily understood.

The Policy Cycle Model

The policy cycle model posits a logical sequence of activities affecting the development of public policies, including environmental policy. It depicts the policy-making process and the broad relationships among policy actors within each stage of it. The model also can be helpful in understanding the flow of events and decisions within different cultures and institutional settings.[1] As indicated in Table 3.1, the model distinguishes six distinct, if not entirely separate, stages in policy making (Jones 1984; see also J. Anderson 1994).

Agenda Setting   In the first stage, a number of activities can be subsumed under what is usually called *agenda setting*. These include the perception and definition of problems, the development of public opinion on them, and the organization of public concerns and new policy ideas to demand action by government. Agenda setting comprises all those activities that bring environmental problems to the attention of the public and political leaders and that shape the ideas and policy alternatives that get serious consideration in government. Because of its critical role in explaining the

Table 3.1   **The Policy-Making Process**

- **Agenda Setting** (perceiving and defining problems)
- **Policy Formulation** (developing proposed courses of action, analyzing policy goals and means)
- **Policy Legitimation** (authorizing or justifying policy action, securing public and political support)
- **Policy Implementation** (putting programs into effect)
- **Policy and Program Evaluation** (assessing policy and program effects, determining success or failure)
- **Policy Change** (revising policy goals or means)

rise of the environmental movement and the evolution of environmental policy, I elaborate on agenda setting below.

Policy Formulation   The formulation of environmental policy involves the development of proposed courses of action to "solve" the problem identified. It includes scientific research on the causes and consequences of environmental problems (including projections of future trends) and review and assessment of the evidence. It requires analysis of policy goals and means in light of economic, technical, political, cultural, ethical, and other considerations. Careful policy design that considers the characteristics of target populations and which tools or incentives will likely bring about the intended behavior is especially important (Schneider and Ingram 1990).

Comprehensive environmental policy analysis would seem a prerequisite for policy formulation. Yet it faces substantial intellectual, political, and institutional barriers in the U.S. political system and elsewhere (Dryzek 1987; Bartlett 1990). The result is that policy formulation typically proceeds incrementally. Existing policies are modified at the margins, in part because it is easier to obtain consensus by sticking with previously agreed upon courses of action. The analytic tasks are also made manageable by considering only a limited and familiar set of policy alternatives and their (largely short-term) consequences (Lindblom 1959). This kind of "muddling through," however, also has important drawbacks. Short-term and narrow assessments of problems and policy alternatives may invite unacceptable long-term costs (Ophuls and Boyan 1992, ch. 6). Alternatives commonly found in the policy literature include what can be called progressive or ambitious incrementalism and trial and error decision making.

Jessica Tuchman Mathews has proposed a strategy of ambitious incrementalism for dealing with international environmental policy at a time of crises and great opportunities. It involves,

> following the path of least resistance; eliminating policies that are both environmentally and economically counterproductive; taking steps that cost little or nothing or those that have immediate economic payoffs; aggressively exploiting existing technology; using well-tested policy instruments, and avoiding the highest political hurdles. It emphasizes the relatively modest steps needed to weave environmental concerns into the fabric of mainstream economic and foreign policy (Mathews 1991a, 310).

Morone and Woodhouse (1986) propose trial and error decision making for handling potentially dangerous technology, such as nuclear power. This strategy involves a sequence of policy moves using trials or experimental efforts, paying attention to feedback from the experience, and then mounting a revised trial that reflects learning from first-round errors. Any number of iterations or cycles might be run as part of a continuous policy-making process. Trial and error decision making simplifies and otherwise focuses complex analysis and choices.

A multiplicity of policy actors play a role in policy formulation, from environmental and business groups to think-tank policy analysts and formal policy makers and their staffs in executive offices and legislatures. They are not equal in their resources or political influence. Indeed, some theorists worry that policy making can be dominated by one set of interests or another (the business community or environmentalists), and thus distort the nation's ability to devise effective and equitable environmental policies (Dryzek 1987; Lindblom and Woodhouse 1993). Some posit that policy specialists (the popular press calls them policy wonks) may dominate, creating the possibility that technocratic decision making triumphs over democracy (see Fischer 1990). Others argue that policy analysts and scientists are not influential *enough* in the process to ensure that rational and technically sound decisions are made (Ophuls and Boyan 1992).

Policy experts within government and outside do play a significant role. Indeed, ad hoc policy task forces or commissions may do much of the work before a proposal is modified and formally endorsed by elected officials. But unlike Western European nations, in the U.S. system elected officials and their appointed top-level assistants rather than permanent professional staff in the agencies make the final policy decisions. The National Energy Strategy that George Bush proposed to Congress in early 1991, for example, followed 18 months of study by a policy task force in the Department of Energy (DOE). The task force held extensive public hearings, consulted closely with other federal agencies, and endorsed strong energy conservation initiatives. In this case, the Bush White House significantly modified the strategy before sending it to Capitol Hill. In particular, Bush's top economic and political advisers persuaded the president to eliminate virtually all of the important energy conservation proposals.

Policy Legitimation   Policy legitimation is usually defined as giving legal force to decisions, or authorizing or justifying policy action—for example, through a majority vote in a legislature, or a formal bureaucratic or judicial decision (Jones 1984). It is much more than that. It refers to the legitimacy of action taken (whether it is viewed as a proper exercise of governmental authority) and the broad acceptability of the action to pertinent publics. Legitimacy or acceptability can flow from several conditions: the action is consistent with constitutional or statutory specifications; it is compatible with U.S. political culture and values, and it has demonstrable popular support; or it has been approved through a process of political interaction of relevant publics and policy officials. That interaction would be expected to involve what some call advocacy and coalition building, including a reasonably full dialogue on the issues and reconciliation of diverse perspectives on what to do about the environmental problem at hand. There is always a risk that some legitimate interests may be excluded from the process, whether by design or through their inability to participate for lack of time, expertise, knowledge of the opportunities, or adequate finances.

The mere passage of legislation or adoption of a regulation is not a sufficient guarantee that policy legitimation has occurred. In some cases, such as the National Environmental Policy Act (NEPA), legislation may be enacted with little serious consideration of the likely impacts. NEPA sailed through the House and Senate in 1969 with virtually no opposition and few asking what it meant. Much the same was true of the demanding Clean Air Act Amendments of 1970 (Jones 1975). Concern over costs and other impacts is so great in the 1990s that such oversights are far less likely to occur today, although rapid action by Congress in 1995 on the "Contract With America" provided few opportunities to consider its effects.

Like policies that are carelessly formulated (e.g., using bad data, questionable projections, or unreasonable assumptions), policies adopted or changed without legitimation run some important risks. They may fail due to technical misjudgments or inaccurate appraisals of public acceptability. Such was the fate of the Nuclear Waste Policy Act of 1982, the authors of which seriously misjudged the public's willingness to accept nuclear waste repositories and its trust and confidence in DOE—the department in charge of the program. Such risks of policy failure may be minimized by ensuring participation by relevant interests or stakeholders (including citizens), opportunities for review and appraisal of proposals, and maintenance of political accountability for decision makers.

Policy Implementation   Policy implementation refers to activities directed toward putting programs into effect. These include interpretation of statutory language, organization of bureaucratic offices and efforts, provision of sufficient resources (money, staff, and expertise), and the details of administration such as provision of benefits, enforcement of environmental regulations, and monitoring of compliance. Those activities occur at all levels in the United States—federal, state, and local—as well as internationally. Most federal environmental protection policies are implemented routinely at the state level through delegation of authority to the states equipped to assume the responsibility.

Implementation is rarely automatic and it involves more than a series of technical and legal decisions by bureaucratic officials. It is deeply affected by political judgments about statutory obligations, priorities for action, provision of resources, and selection of implementation tools. It is also influenced by the responses or expected responses of target groups and other publics. Variables such as the commitment and administrative skills of public officials in charge of the program make a difference as well (Mazmanian and Sabatier 1983).

Policy and Program Evaluation   Once implemented, we need to ask whether environmental policies and programs are working well or not. We judge the merit of the decision-making processes and program achievements against some specified standard, such as efficiency of resource use or effectiveness in achieving stated goals and objectives (Knaap, Kim, and Fitipaldi 1996). There are several different ways in which environmental policies may be evaluated (see Chapter 6), but the most common is to ask whether they produce expected outcomes. For example, does the Clean Air Act result in cleaner air? Evaluations may be rigorous attempts to measure and analyze specific program outcomes and other effects and to relate them causally to policy efforts. However, they may also be far less systematic investigations by congressional oversight committees, internal agency review bodies, or interest groups. As is true of all stages in the policy cycle, political pressures and judgments affect whether, and to what extent, policy makers consider such information.

Policy Change   The last stage in the cycle is policy change. If the outcomes are not satisfactory, environmental policies may be revised in an attempt to make them more successful, or they may be terminated. New policy goals may be proposed, new authority granted to an agency, more funds appropriated for programs, new approaches (such as using market-based incentives) endorsed, or priorities for implementation changed. Termination of government programs is rare, but environmentalists have suggested taking exactly that action for many natural resource policies that have out-lived their original purposes or which have resulted in unacceptable degradation of the environment. Examples include many western land and water use policies stoutly defended by politically powerful constituencies that benefit from their con-tinuation.

Although analytically distinct and logically arranged, this sequence of activities in the policy process may follow a different order, the stages may overlap one an-other, and they may take place in more than one institutional setting. Policy actors come and go, new data and arguments are advanced, problems are defined and rede-fined, and policy solutions proffered and judged.

Patterns in Environmental Policy Making

A general description such as this one of the policy-making process also conveys lit-tle of the high stakes involved in environmental policy decisions, the role of power-ful interest groups and individuals, the opportunities for citizen influence, and the unpredictability of the policy results. Policy making is not tidy, nor does it always meet expectations for either democracy or ecological rationality. Dean Mann ob-served almost ten years ago that the outcomes of the process fall short of conven-tional models of problem solving in part because of the character of the U.S. politi-cal system:

> That the politics of environmental policymaking is a process of dramatic advances, incomplete movement in the 'right' direction, frequent and partial retrogression, sometimes illogical and contradictory combina-tions of policies, and often excessive cost should come as no surprise to students of American politics (Mann 1986, 4).

If the policy results are not good enough, we need to ask what might be changed in the institutional arrangements of American government or in its politics to get closer to where the nation needs to be. Studies of environmental and other regulatory pol-icy making offer some insights.

As might be expected, the pattern of policy making varies significantly from one problem area to another, which suggests that governmental structure is not the only factor that matters. The politics of western water use is different from the politics of controlling toxic chemicals and air pollution. Ingram and Mann (1983), for example, argue that the types of environmental policies adopted (e.g., regulatory or distribu-tive), reflect political variables such as the structure of demand (conflict or consen-sus among interest groups), the structure of decision making (integrated or frag-mented government institutions), and the structure of impacts (the actual effects on society of the policies, including costs and other burdens and on whom they fall).

Other important variables that help to explain why we get the environmental and natural resource policies we do include the perception of policy impacts, particularly the concentration or dispersal of costs and benefits and hence the incentives that are created for different actors to participate in the policy-making process (Wilson 1980). Narrow economic interests (automobile manufacturers, ranchers, loggers, and mining companies) adversely affected by proposed environmental policies, for example, have a reason to organize and fight them. The public receiving the broadly dispersed benefits of environmental protection is not usually so stimulated to rise in their defense. Yet what Wilson calls "entrepreneurial politics" may alter the usual logic of collective action (Olson 1971), where the public has little material incentive to organize or actively support actions that benefit society as a whole. Policy entrepreneurs in environmental groups and Congress mobilize latent public sentiment on the issues, capitalize on well-publicized crises and other catalytic events, attack their opponents for endangering the public's welfare, and associate proposed legislation with widely shared values (clean air, public health). When the benefits of environmental policies flow to narrow interests (e.g., ranchers, farmers, miners, or loggers), however, and the costs are broadly distributed, the beneficiaries are likely to organize and lobby to protect their interests while the general public will usually prove hard to mobilize against natural resource subsidies or "giveaways" that are low salience issues for most people. Whether we get what Wilson calls "client politics" under such circumstances depends on how visible the policies are and the extent to which the public and environmental groups are able to challenge the beneficiaries effectively.

As these examples illustrate, the salience and complexity of the issues and the degree of conflict that exists over them are important factors (Gormley 1989). These qualities affect whether the public and policy makers take a strong interest and choose to get involved or not. Elected officials may prefer not to engage complex issues of low salience and high conflict because they consume valuable time to acquire the necessary expertise. There also are few political benefits given the low visibility of the issues and significant political risks due to conflict among opposing interests. Thus politicians may ignore the issues altogether or allow those with expertise and strong interest in the outcome to dominate the process.

These theoretical observations hint at the pivotal role played in policy making by environmental and other public interest groups that arose during the 1960s and 1970s, and the policy entrepreneurs who championed their causes. They also suggest that broad support for environmental policy by the American public has been the greatest source of political power for those groups. These forces can best be understood as components of the broader process of environmental agenda setting.

AGENDA SETTING AND ENVIRONMENTAL POLICY CHANGE

In one effort to explain the rise and fall of issues on both societal and governmental agendas, John Kingdon (1984) proposed an intriguing model. It is useful for understanding how environmental problems come to be objects of public concern or not, how they gain or fail to gain the attention of public officials, and how environmental policy takes the form it does. The model also serves as a guide to developing political strategies for influencing the future development of environmental policy.

Problems, Policies, and Politics

Kingdon argues that three separate but interdependent "streams" of activities (related to problems, politics, and policies) flow continuously through the political system. They sometimes converge, with the assistance of policy entrepreneurs or political leaders, and create windows of opportunities for policy development. Much of this activity occurs within policy communities of specialists and interested parties, such as environmental and industry groups that focus on air and water pollution, pesticide use, energy use, or population issues.

Kingdon defines the agenda as "the list of subjects or problems to which government officials, and people outside of government closely associated with those officials, are paying some serious attention at any given time" (Kingdon 1984, 3). The governmental agenda may be influenced by the larger societal agenda, the problems that most concern people, as expected in a democracy when the public is mobilized around salient issues. It may also be shaped by the diffusion of ideas among policy communities (or policy elites) or by a change in the political climate, as occurred with the election of Ronald Reagan in 1980 and Bill Clinton in 1992. The way in which the government's agenda is set depends on the flow of those problem, policy, and politics streams.

The *problem stream* influences the agenda by providing data about the state of environmental conditions and trends, as reviewed in Chapter 2. The information may come from government reports and program evaluations such as EPA studies of air or water quality, assessments by the National Academy of Sciences and other scientific bodies, reports by presidential commissions, and studies sponsored by environmental groups, industry, and others. The data and assessments circulate among policy specialists, affecting their perceptions and understanding of the problems regardless of whether they produce any immediate effects on policy decisions. The problem stream is also affected by exogenous variables such as environmental crises or disasters, technological developments, and ecological changes. Accidents such as the chemical plant explosion at Bhopal, India, in 1984 and the Exxon *Valdez* oil spill in Alaska in 1989 often receive extensive media coverage. Such reporting may prompt people to pay greater attention to the problems of dangerous chemicals and transportation of oil in vulnerable supertankers. Catalytic events like these accidents increase the credibility of studies and reports that document the environmental problem at issue, and they help to ensure they will be read and debated, and thus influence policy decisions.

The *policy stream* concerns what might be done about environmental problems. Proposals are developed by any number of analysts, academics, legislators, staffers, and other policy actors, as noted under policy formulation. They are floated as trial balloons and become the objects of political speeches, legislative hearings, and task forces. They get tested by the policy community for technical acceptability and political and economic feasibility. They are endorsed or rejected, revised, and combined in new ways, somewhat like a process of biological natural selection. Ideas that are inconsistent with the current political mood may be dropped from consideration and relegated to the policy back burner for warming or incubation until the climate improves. Such was the fate in the early 1990s of proposed stiff carbon taxes to discourage consumption of fossil fuels. Conventional policy alternatives such as "command and control" regulation may drop from favor while other ideas, such as market-based incentives, are viewed more positively. The language used and symbols evoked can make the difference between acceptance and rejection, as can changes in our understanding of the problems. Policy actors may avoid population "control" policies as coercive while they embrace voluntary family planning as consistent with cultural values of individual choice. In this way, a short list of acceptable policy alternatives emerges at any given time.

Finally, the *politics stream* refers to the political climate or national mood as revealed in public opinion surveys, election results (particularly a change in presidential administrations), and the activities and strength of interest groups. The political mood is never easy to decipher, and sometimes judgments are well off the mark, as was the case with the reputed Reagan election mandate in 1980. Many Reagan supporters and political analysts assumed incorrectly that the public became more conservative on environmental issues during the 1980s (Kraft 1984; Dunlap 1987). Yet most elected officials develop a well-honed ability to detect important shifts in public attitudes, at least in their own constituencies. That is why environmentalists and other advocacy groups try to mobilize the public around their issues by stimulating a sense of public outrage over existing problems or actions by policy makers with which they disagree.

Policy Entrepreneurs and Policy Change

Although these three streams flow independently, at times they come together. Environmental policy entrepreneurs, leaders inside and outside of government who devote themselves to the issues and their advancement, help to bring this convergence about. They especially do so when they see windows of opportunity, such as a major accident or crisis or the beginning of a new presidential administration. After over ten years of congressional inaction on oil-spill legislation, the Exxon *Valdez* spill prompted Congress to enact the Oil Pollution Prevention Act of 1990. It required companies to submit oil-spill contingency plans to the Coast Guard and EPA and to train their employees in oil-spill response.

Policy entrepreneurs are prepared to take advantage of such opportunities. In the meantime, they continue to stimulate interest in the problems, educate the public and policy makers, circulate new studies, and otherwise incubate the issues.[2] They are not equal in their ability to perform those essential tasks. Indeed, critics of U.S. politics often suggest important differentials in political access and influence that some term a "mobilization of bias." Environmental and other public interest groups have greatly increased their political clout over the past 25 years. Nonetheless, according to several recent surveys, they still lack the financial and other resources common among business and industry groups (Furlong 1992; Schlozman and Tierney 1986).

This kind of convergence of the three streams helps to explain some peculiar patterns of environmental attention and inattention. Energy issues, for example, were at the top of the political agenda in the late 1970s as President Jimmy Carter sought (but failed) to enact a comprehensive national energy policy. However, they disappeared from sight in the 1980s as the White House and members of Congress lost interest in the subject when energy prices fell and public concern dissipated. Attention increased again in 1988 as a hot and dry summer stirred fears of global warming. That concern was aided by the activism of scientists such as James E. Hansen, director of NASA's Goddard Institute for Space Studies, and the noted climatologist Stephen Schneider. They spoke out frequently (unusual for scientists) about the risks of climate change and the need for governmental action.

Critics of U.S. energy policy argue that this cycle from crisis to complacency and back ill serves the nation as much as it is understandable in the dynamics of agenda setting. Much the same is true of other major environmental or natural resource problems (for example, biodiversity and population growth) that typically are low salience issues even if from time to time people view them as "crises" demanding immediate action. The challenge to environmental and other groups concerned about those issues is to raise their salience through public education and political activism, and to develop the skills of the successful policy entrepreneur that help to translate scientific knowledge and diffuse public concern about environmental problems into public policy.

These kinds of agenda-setting activities also account for transformations in environmental policy over time. Christopher Bosso (1987) studied changes in federal pesticide policy from the 1940s to the 1980s, and attributed the shift largely to challenges by environmental and consumer groups to the once dominant agricultural chemical industry and its allies in Congress and the Agriculture Department. The older "politics of clientelism" and its reigning paradigm based on the benefits of pesticide use, Bosso argues, collided head on in the 1960s and 1970s with new concerns for safety and environmental quality advanced by environmentalists and others. In short, the definition of the problem was altered, as were the implications for public policy. The change occurred because new public attitudes about chemical risks were emerging, and environmental advocacy groups were able to advance the case for regulation in a Congress that by the 1970s increasingly reflected such new public concerns. The result has been a gradual shift away from the old pesticide policy, with frequent bouts of policy gridlock as the two coalitions of interests do battle in Congress and in the agencies.

Characteristics of U.S. Government and Politics

Some unique characteristics of the U.S. political system shape the policy process outlined here and the environmental policies that emerge from it.[3] Formal institutional structures, rules, and procedures are never neutral in their effects. As Schattschneider (1960, 71) observed 35 years ago, all political organizations "have a bias in favor of the exploitation of some kinds of conflict and the suppression of others because *organization is the mobilization of bias.* Some issues are organized into politics and others are organized out." We study institutions because they channel political conflict and thus affect the policy process and its results.

Constitutional and Political Features

The U.S. Constitution sets out the basic governmental structure and establishes an array of individual rights that have been largely unchanged for over 200 years.

Government authority is divided among the three branches of the federal government and shared with the 50 states, and some 80,000 local units of government. The logic of the tripartite arrangement of the federal government was to limit its authority through creation of separate and countervailing powers in each branch and to protect individual rights (to prevent coercion or tyranny). Additional guarantees of freedom for individuals (and corporations) were provided in various sections of the Constitution. This is most notably the case in the due process clause of the Fifth Amendment, which put a premium on protection of property rights and thereby created significant barriers to governmental action.[4] Decentralization of authority to the states likewise reflected public distrust of the national government in the late eighteenth century and a preference for local autonomy. At that time, the nation's small population of four million lived largely in small towns and rural areas, and the activities of the federal government were minuscule compared to its present size and scope of responsibilities. Yet the constraints placed on government authority to act, on majority rule, and on prompt policy development continue in the late twentieth century.

Other constitutional dictates and political influences also have important implications for environmental policy. These include staggered terms of office for the president, senators, and representatives, which tend to make members of Congress independent of the White House. That proclivity is reinforced by the geographic basis of representation and an electoral process that induces members to pay more attention to local and regional interests directly related to their reelection than to the national concerns that preoccupy presidents and executive branch officials. To this inherent legislative parochialism, we can add a preoccupation with individual political goals. Members of Congress have assumed almost complete responsibility for their own political fundraising and reelection campaigns as political parties have become enfeebled and the growth in the number and political influence of narrowly focused interest groups (Berry 1989) has replaced the broader integrative forces of parties and the presidency.

Competition between the two major parties also inhibits coalition building and the development of comprehensive and coordinated environmental policies. An extensive scholarly literature confirms a strong association between partisanship and environmental policy support among elected officials. Democrats are far more supportive of environmental protection policy than are Republicans (Calvert 1989). The differences are clearly evident in congressional voting scores compiled by the League of Conservation Voters (LCV) and in most years are also seen in party platforms (e.g., Kraft 1984). In 1993, for example, Democrats averaged a 71 percent LCV score in the House while Republicans averaged 32 percent; the parallel figures for the Senate were 70 percent and 16 percent (League of Conservation Voters 1994). There is evidence that Republican voters are becoming more like Democratic voters; the Republican party platform in 1992 also reflected a greater sensitivity to the environment. Nevertheless, the very strong differences between Democrats and Republicans on environmental voting in Congress remain. The explanation for the divergence of Republican voters and elected Republican officials may lie in the low salience of environmental issues and the need for

Republicans to respond to the business community, an important component of their constituency (Kamieniecki 1995).

Institutional Fragmentation and Policy Gridlock

These formal constitutional and informal political forces have led some scholars to question whether the U.S. government is capable of responding in a timely and coherent way to environmental challenges (Ophuls and Boyan 1992). Constitutionally created checks and balances may constrain abuses of authority, but they generate the opportunity for policy gridlock as hundreds of environmental and industry groups struggle to block each other in the often labyrinthine corridors of power in Congress, executive branch agencies, and the federal courts. When issues are less visible and contentious, the same groups may have an inordinate influence on environmental policy because of their easy access to key policy makers and their valued expertise.

Dispersal of Power in Congress   The congressional committee system is a good example of the tendency to fragment authority in the U.S. political system. Seven major committees in the House and five in the Senate are responsible for environmental policy. Each of those committees also has several semiautonomous subcommittees (see Table 3.2). The committees often disagree on policy actions, and they depend on each other in crucial ways. For instance, the authorizing committees that establish environmental programs rely on the appropriation committees to supply the money to run them.[5]

In some respects, the dispersal of power in Congress is even greater than suggested by looking only at the activities of the major environmental committees. For example, the EPA reported in 1987 that 14 committees and 20 subcommittees in the Senate and 18 committees and 38 subcommittees in the House had *some* jurisdiction over the agency's activities (U.S. Environmental Protection Agency 1987a). Despite this obvious fragmentation of authority, several long-term committee chairs in Congress have been extraordinarily influential in recent years in determining the direction of environmental and energy policy. These include Representative John Dingell (D-Michigan), formerly chair of the powerful House Energy and Commerce Committee, and Senator J. Bennett Johnston (D-Louisiana), formerly chair of the Senate Energy and Natural Resources Committee. Upon winning control of the House in 1995, the Republicans eliminated one environmental committee, Merchant Marine and Fisheries, and slightly altered the jurisdictions (and names) of several others. However, fragmentation of power remained largely unchanged.

One effect of the dispersion of power in Congress is especially important. Building consensus on policy goals and means is often unattainable because of the diverse policy actors and the multiplicity of committees involved (Kraft 1995a). Action may be blocked even when public concern about the environment is high and consensus exists on broad policy directions. The reason is that it is much easier to stop legislative proposals from going forward than it is to build far-reaching environmental policies acceptable to all parties. Moreover, interest groups have myriad ways to

Table 3.2 Leading Congressional Committees with Environmental Responsibilities

Committee	Environmental Policy Jurisdiction
House of Representatives	
Agriculture	Agriculture in general; soil conservation; groundwater; forestry and private forest reserves; pesticides and food safety.
Appropriations[a]	Appropriations for all programs.
Commerce (formerly Energy and Commerce)	Air pollution; national energy policy in general; exploration, production, pricing, and regulation of energy sources; nuclear energy and nuclear waste; energy conservation; safe drinking water; superfund and hazardous waste disposal, toxic substances control; noise control.
Resources (formerly Natural Resources)	Public lands and natural resources in general; national parks, forests, and wilderness areas; irrigation and reclamation; mines and mining; energy, and nuclear waste disposal; oceanography, fisheries, international fishing agreements, and coastal zone management; and wildlife, marine mammals, and endangered species.
Transportation and Infrastructure (formerly Public Works and Transportation)	Water pollution, flood control; rivers and harbors; pollution of navigable waters; dams and hydroelectric power; transportation; Superfund and hazardous wastes.
Science (formerly Science, Space and Technology)	Environmental research and development; energy research and development; research in national laboratories; global climate change; National Aeronautics and Space Administration, National Oceanic and Atmospheric Administration, and National Science Foundation.
Senate	
Agriculture, Nutrition and Forestry	Agriculture in general; soil conservation and groundwater; forestry and private forest reserves; pesticides and food safety; global change.
Appropriations[a]	Appropriations for all programs.
Commerce, Science and Transportation	Coastal zone management; inland waterways, marine fisheries; oceans, weather, and atmospheric activities; Outer continental shelf lands; technology research and development; surface transportation; global change.
Energy and Natural Resources	Energy policy in general; conservation, research and development; oil and gas production and distribution; nuclear waste policy; mining; national parks and recreation areas; wilderness; wild and scenic rivers; public lands and forests, global change.

Table 3.2 *Continued*

Committee	Environmental Policy Jurisdiction
Environment and Public Works	Environmental policy and research in general; air, water and noise pollution; safe drinking water, environmental aspects of outer continental shelf and ocean dumping; toxic substances other than pesticides; Superfund and hazardous wastes; solid waste disposal and recycling; nuclear waste policy; fisheries and wildlife; flood control and dams, and improvements of rivers and harbors; water resources; global change.

Source: Compiled from descriptions of committee jurisdictions reported in *Players, Politics, and Turf of the 103rd Congress* (Washington: Congressional Quarterly Inc., May 5, 1994). Committee jurisdictions and name changes in the 104th Congress are drawn from reports in *Congressional Quarterly Weekly Report,* particularly the issues of December 17, 1994 and January 21, 1995.

ᵃ Both the House and Senate Appropriations Committees have Interior subcommittees that handle all Interior Department agencies as well as the Forest Service. In both houses a subcommittee on VA, HUD, and Independent Agencies is responsible for EPA appropriations. The Energy Department, Army Corps of Engineers and Nuclear Regulatory Commission fall under the jurisdiction of the subcommittees on Energy and Water Development. Tax policy affects many environmental, energy, and natural resources policies, and is governed by the Senate Finance Committee and the House Ways and Means Committee.

shape legislative proposals they cannot halt so that damage (from their perspective) is minimized.

Divided Authority in the Executive Branch   Similar institutional pluralism characterizes the executive branch, as illustrated in Figure 3.1. The EPA has chief responsibility for the major environmental statutes, and its work is divided among four separate offices dealing with specific environmental media such as air and water, discouraging cross-media and other forms of integrated environmental decision making (see Chapter 4).

In addition to the EPA, 11 cabinet departments have significant roles in environmental policy. Four departments have major responsibilities for environmental protection and natural resources: Interior, Agriculture, Energy, and State (the last for international policies). Independent agencies such as the Nuclear Regulatory Commission and selected offices in the executive office of the president (the CEQ, Office of Environmental Policy, Office of Management and Budget, Council of Economic Advisors, Office of Science and Technology Policy, and the new National Science and Technology Council) are also regular participants in formulating and implementing environmental policies.

Under the conditions prevailing in both Congress and the executive branch, policy making depends on bargaining among power wielders to frame temporary compromises most can live with. In turn, this means that environmental policies (like other public policies) ordinarily change incrementally (in small steps and slowly) rather than dramatically or quickly, although exceptions occur. Successful policy

President

The Executive Office of the President

White House Office	Council on Environmental Quality	Office of Management and Budget	Office of Science and Technology Policy
Overall policy	Environmental policy coordination	Budget	Advises president on issues involving science, technology and engineering
Agency coordination	Oversight of the National Environmental Policy Act	Agency coordination and management	
Office of Environmental Policy	Environmental quality reporting		

Environmental Protection Agency	Department of the Interior	Department of Agriculture	Department of Commerce	Department of State
Air & water pollution	Public lands	Forestry	Oceanic and atmospheric monitoring and research	International environment
Pesticides	Energy	Soil conservation	Coastal zone management	Population
Radiation	Minerals		Marine mammal protection	Development assistance
Solid waste	National parks			
Superfund	Wilderness			
Toxic substances	Wildlife			
	Endangered species			
	Continental shelf			

Department of Justice	Department of Defense	Department of Energy	Department of Transportation	Department of Housing and Urban Development
Environmental litigation	Civil works construction	Energy policy coordination	Mass transit	Housing
	Dredge & fill permits	Nuclear waste disposal	Roads	Urban parks
	Pollution control for defense facilities	R & D	Airplane noise	Urban planning
	Environmental cleanup and restoration	Waste management	Oil pollution	
		Environmental restoration		

Department of Health and Human Services	Department of Labor	Nuclear Regulatory Commission	Tennessee Valley Authority
Health	Occupational health	Licensing and regulating nuclear power	Electric power generation
Family planning			

Figure 3.1   Executive Branch Agencies with Environmental Responsibilities

Sources: Council on Environmental Quality, *Environmental Quality: Sixteenth Annual Report of the Council on Environmental Quality* (Washington, D.C.: U.S. Government Printing Office, 1987), *United States Government Manual 1993/94* (U.S. Government Printing Office, 1993), and author.

making also requires skillful and determined political leadership capable of assembling ad hoc coalitions, fashioning legislative and executive compromises, and shepherding the resultant measures through Congress or an executive agency. Senate Majority Leader George Mitchell's (D-Maine) leadership on the Clean Air Act of 1990 exemplified those qualities (Cohen 1992).

The Effects of Dispersed Power

Would more centralized or integrated political institutions produce a better outcome than the decentralized and loose policy making apparatus we now have? The answer is not clear. Despite criticism of the present institutional arrangements in government, a decentralized and competitive political process has some attractive and often overlooked qualities. Among them are the many opportunities created for interest groups and policy entrepreneurs to promote issues of concern to them. There are also innumerable points of access in the permeable U.S. system for those who wish to oppose the policies promoted by the president or members of Congress.

The extreme decentralization of the congressional committee system, for example, virtually guarantees that environmental advocacy groups (or their opponents) can find a friendly audience somewhere to publicize their cause. Environmentalists discovered the attractiveness of this kind of "fragmentation" during Ronald Reagan's presidency (1981 to 1989) when they became highly adept at stimulating public and congressional opposition to the president's efforts to weaken environmental policies (Kenski and Kenski 1984). Between May 1981 (when she was confirmed as administrator of the EPA) and July 1982, Anne M. Gorsuch (later Burford) was forced to testify 15 times before congressional committees. Other EPA officials appeared more than 70 times between October 1981 and July 1982, and between 56 and 79 times each year from 1984 to 1986 (U.S. Environmental Protection Agency 1987b).

Much the same argument applies to the courts. Even before Reagan assumed office, environmentalists had come to rely heavily on using the federal courts to pressure reluctant executive agencies to implement the tough new statutes adopted in the 1970s. They continue to do so where their resources allow (Wenner 1994). Much of the bitter fight over protection of old growth forests in the Pacific Northwest has taken place in the federal courts, with either environmentalists or the logging industry challenging administrative plans that sought to balance competing interests (Yaffee 1994).

In much the same way, environmentalists have turned to state and local governments when Washington proved unwilling to respond to their agenda. Some of the most innovative environmental policies over the past decade have come not from the federal government but from the "laboratories of democracy," the states (Lowry 1992; John 1994). Revisions in federal environmental law often draw from experiments in at least some of the states, as Congress did in incorporating into the 1990 Clean Air Act some of California's initiatives on air pollution. Wisconsin's strict policy regulating the use of CFCs in automobile air conditioners influenced ozone-protection provisions of the same 1990 act. Local governments, particularly in progressive communities and university towns, such as Davis, California, have adopted their own distinctive policies on everything from mandatory recycling programs and use of alternative energy sources to climate change. State and local actions of this kind have created a richness and diversity in environmental policy that would not be possible in a more centralized and unified system.

These characteristics suggest the need for judicious appraisal of institutional structure and performance. Critics are correct to note that the U.S. political system suffers from serious institutional deficiencies when weighed against the imperatives of contemporary environmental policy needs, or for that matter against the need to gain firmer control over a range of other government programs. U.S. politics also has important strengths that environmentalists and other political activists know quite well. They benefit from them in much of what they do to shape public policy.

Chapters 4 through 6 discuss the way U.S. government and politics affect environmental decisions. Another way to judge governmental performance is to take a brief retrospective look at the development of environmental policy. Such an examination offers persuasive evidence that government has been fairly responsive to changing public concerns about the environment, particularly when the issues are politically salient. Public opinion has been a driving force in the shaping of modern environmental policy. What is less clear is whether that pattern will continue in an era when many environmental policy decisions are increasingly technical, made largely in administrative agencies and the courts rather than in legislatures, and few who are not active participants in Washington-based issue networks understand them.

THE EVOLUTION OF NATURAL RESOURCE AND ENVIRONMENTAL POLICIES

Modern environmental policy emerged in the 1960s and became firmly established on the political agenda during the "environmental decade" of the 1970s when Congress enacted most of the contemporary environmental statutes. Wisconsin's Senator Gaylord Nelson organized the first Earth Day held on April 22, 1970. It was widely celebrated across the nation and signalled the arrival of a mass political movement dedicated to ending environmental degradation. Politicians responded eagerly to what many considered a motherhood issue that posed little political risk and many electoral dividends.

Yet concern about the environment and the value of natural resources rose early in the nation's history and periodically sparked the adoption of preservation, conservation, economic development, and public health policies that continue to shape the nation's environmental agenda today. Most of those policy actions coincided with three periods of progressive government in which social and economic regulation also advanced: the Progressive Era from about 1890 to 1915, the New Deal of the 1930s, and the expansion of "social regulation" in the 1960s and 1970s. In each period, perceived crises, catalytic events, and the leadership of policy entrepreneurs heightened public concern and kindled policy innovation. Broader social, economic, and technological changes also contributed to recognition of looming environmental and resource problems and created the political will to deal with them. These developments helped to set the intellectual, scientific, aesthetic, moral, and political foundations of contemporary environmentalism (Shabecoff 1993).[6]

Settlement and "Conquest" of Nature

In the early seventeenth century, in one of the first conservation actions in the new nation, New England colonists adopted local ordinances protecting forest land and regulating timber harvesting (Nash 1990). Such actions did not prevent the colonists from attempting to subdue the wilderness they faced and fundamentally altering the landscape and ecology of New England as population grew and the land was cleared for agriculture and settlements (Cronon 1983). That behavior was an early indication of the limits of policy intervention in the face of profound pressures for expansion and a culture that favored exploitation of the nation's rich natural resources.

By the middle of the nineteenth century, new trends were emerging. The study of natural systems was gaining stature within the scientific community, and in 1864 George Perkins Marsh published his influential treatise *Man and Nature,* documenting the destructive impact on nature from human activity. Literary figures contributed as well to the sense that industrialism and technology were not entirely beneficent in their effects. Henry David Thoreau's *Walden,* a poignant account of his two years in the wilderness at Walden Pond, reflected many of the same concerns as today's "deep ecology" writing. The late nineteenth century brought some advances in conservation policy as the consequences of human activities began to attract more attention, although by present standards the effects of the policies were modest. Discoveries of vast areas of unsurpassed beauty in the newly explored West led to the establishment of Yosemite Valley, California, as first a state park (1864) and later as a national park (1891). In 1872 Congress set aside two million acres in Wyoming, Montana, and Idaho to create Yellowstone National Park as a "pleasuring ground for the benefit and enjoyment of the people," the first of a series of national parks.

During this same period, however, the federal government sought to encourage rapid development of its vast holdings in the West. It adopted public policies toward that end that reflected prevailing beliefs in Congress and elsewhere that the West was an immense frontier promising "limitless opportunities" for resource exploitation and creation of wealth. Chief among those actions were classic distributive policies: the generous distribution of public lands (free or at token prices) to private parties, such as railroad companies and homestead settlers. The Homestead Act of 1862 allowed individuals to acquire 160 acres of public land by living on it and working it as a farm. Over 250 million acres were converted to farms through this act. The federal government transferred over 94 million acres to railroad corporations and an additional 800 million acres to the states, veterans, and other groups under various programs. Over one billion acres of the 1.8 billion in the original public domain were privatized between 1781 and 1977, leaving about 740 million acres of public land, of which 330 million acres are in Alaska (Wengert 1994).[7] Inducements to develop the West also included what in retrospect seem extravagant subsidies to graze animals and mine minerals on public lands, and to farm arid and semiarid lands with the assistance of public irrigation projects. The 1872 General Mining Law, for example, gave miners virtually free access to rich mineral deposits on public land with no obligation to pay royalties on the minerals extracted. The Reclamation Act of 1902 provided for public construction of dams and other projects to make cultivation of desert lands possible. The West gained population and the economy prospered, but at a high cost to the environment.

Another important effect was more political. The unfettered and heavily subsidized use of public lands for such a long period created the belief among recipients that they had a right to receive such benefits indefinitely. The legacy affects natural resource policy even today. For years, powerful western constituencies have dominated these natural resource policies, forming protectionist subgovernments or subsystems in association with members of key congressional committees and executive agencies. Such alliances operate autonomously with little political visibility. Conflict is minimal and easily resolved through logrolling, or mutually beneficial bargaining in which each party may gain its goals. Moreover, participants are cohesive and generally in agreement on policy goals (McCool 1990). These political arrangements allowed the natural resource subgovernments to ward off significant policy changes sought by conservationists and others until very recently (Foss 1960; McConnell 1966; Lowi 1979).

The Conservation Movement and Advances in Public Health

By the late nineteenth century, winds of change began to blow as the Progressive Era unfolded. Reflecting the growth of concern about preservation of natural resources and public lands in response to reckless exploitation in earlier decades, in 1892 John Muir founded the Sierra Club, the first broad-based environmental organization. Muir led the preservationist wing of the incipient environmental movement, with a philosophy of protecting wilderness areas, such as his beloved Yosemite Valley, from economic development. Such areas, Muir argued, should be preserved for their own sake and used exclusively for recreational and educational purposes.

A countervailing conservationist philosophy took hold under Gifford Pinchot. The Yale graduate trained in forestry in Germany and became the first professional forester in the United States. In 1898, Pinchot, who emphasized efficient use (or "wise management") of natural resources for economic development, became chief of the Division of Forestry, the forerunner of the modern U.S. Forest Service. Pinchot's approach to conservation soon became the dominant force in twentieth century natural resource policy in part through his close association with President Theodore Roosevelt. In 1908 Pinchot chaired a White House Conference on Resource Management that firmly established his brand of conservation as the nation's approach to natural resources. The heritage can be seen in key doctrines of what historian Samuel P. Hays (1959) called the "progressive conservation movement," among them "multiple use" and "sustained yield" of the nation's resources.

Despite the differences between Muir and Pinchot, the conservation movement achieved early successes in the creation of national parks and forest reserves, national monuments such as the Grand Canyon (1908) and government agencies such as the Forest Service (1905), National Park Service (1916), and Bureau of Reclamation (1902). Many of the prominent national conservation groups also emerged at this time. In addition to the Sierra Club, the National Audubon Society was founded in 1905 and the National Parks and Conservation Association in 1919. Other environmental organizations emerged between the World Wars I and II, including the Izaak Walton League in 1922, the Wilderness Society (founded by ecologist and writer Aldo Leopold) in 1935, and the National Wildlife Federation in 1936. None of these developments posed a serious challenge to prevailing values regarding the sanctity of private property, individual rights, and economic growth. Rather they represented a somewhat inconsistent and awkward accommodation of new social forces in twentieth century America that eventually would provoke a more spirited opposition. The actions taken, however, did establish the important principle that resources in the public domain should be used equitably for the benefit of all citizens.

Environmental protection efforts focusing on public health were not to appear for the most part until the 1960s. However, there was a Progressive Era parallel to the conservation movement in the establishment of urban services such as wastewater treatment, waste management, and provision of clean water supplies, along with broader societal improvements in nutrition, hygiene, and medical services. Such gains were spurred by concerns that developed in the late nineteenth century over the excesses of the Industrial Revolution and a system of private property rights that operated with virtually none of the restraints common in the 1990s. The first air pollution statutes in the United States, for example, were designed to control heavy smoke and soot from furnaces and locomotives. They were approved in Chicago and Cincinnati in the 1880s, and by 1920, some 40 cities had adopted air pollution control laws. Although a few states such as Ohio took action as early as the 1890s, no *comprehensive* state air pollution policies existed until 1952, when Oregon adopted such legislation (Portney 1990b; Ringquist 1993).

From the New Deal to the Environmental Movement

The environmental agenda during President Franklin Roosevelt's 12-year tenure emphasized the mitigation of natural resource problems, particularly flood control and soil conservation, in response to a series of natural disasters. The most memorable of these was prolonged drought and erosion in the Dust Bowl. Activities such as creation in 1933 of the Tennessee Valley Authority (TVA) were intended to stimulate economic development and employment to pull the nation out of the Depression. Yet the TVA also was important for demonstrating that government land use planning could be used to benefit the broad public in a region. Other actions—for example, the establishment of the Civilian Conservation Corps and the Soil Conservation Service—were directed at repairing environmental damage and preventing its recurrence. Notable among the many New Deal policies was the Taylor Grazing Act of 1934, which was intended to end the abuse from overgrazing of valuable rangelands and watersheds in the West by authorizing the Department of the Interior to issue grazing permits and regulate rangeland use. Congress created the Bureau of Land Management in 1946 and ended the massive privatization of public lands (Wengert 1994).

During the 1950s and 1960s, a third wave of conservation focused on preservation of areas of natural beauty and wilderness, stimulated in part by increased public interest in recreation, and efforts to stem the tide of economic development threatening those areas. The Wilderness Act of 1964 was intended to preserve some national forest lands in their natural condition through a National Wilderness Preservation

System. The Land and Water Conservation Fund Act, enacted in 1964, facilitated local, state, and federal acquisition and development of lands for parks and open
spaces. In addition, Congress created the National Wild and Scenic Rivers System in
1968. At the request of President Lyndon B. Johnson, his wife Lady Bird Johnson,
and the President's Commission on Natural Beauty, Congress approved legislation
to "beautify" federal highways by reducing the number of unsightly billboards
among other actions to promote aesthetic values. Stewart Udall, Secretary of the Interior under both President John F. Kennedy and Johnson, was a forceful advocate of
conservation policies.

This latest "wilderness movement" eventually evolved into the modern environmental movement, with a much broader policy agenda. Congress adopted the first
federal water and air pollution control laws in 1948 and 1955, respectively, and
gradually expanded and strengthened them. The federal government gingerly approved its first international population policies in the early 1960s several years after
congressional policy entrepreneurs held hearings and incubated the controversial
measures. In his 1965 State of the Union address, President Johnson called for federal programs to deal with "the explosion in world population and the growing
scarcity in world resources." The following year Congress authorized the first funds
for family planning programs abroad (Kraft 1994b).

The successes of conservation efforts and the nascent environmental movement
were fundamentally dependent on long-term changes in social values that began after World War II and accelerated as the nation developed economically and the production and consumption of consumer goods escalated dramatically in the 1950s
and 1960s. The United States slowly shifted from an industrial to a postindustrial or
postmaterialist society. An increasingly affluent, comfortable, and well-educated
public placed new emphasis on the quality of life (Hays 1987; Inglehart 1990). Concern for natural resource amenities and environmental protection issues was an inte-
gral part of this change, and by the 1970s it was evident across all groups in the population, if not to the same degree. Similar factors help to account for the growth of
the global environmental movement at the same time (McCormick 1989).

Scientific discoveries brought new attention to the effects of pesticides and other
synthetic chemicals on human health and the natural environment. These were documented in Rachel Carson's influential *Silent Spring* (1962), Murray Bookchin's *Our
Synthetic Environment* (published about six months before Carson's book under the
pseudonym of Lewis Herber), and Barry Commoner's *The Closing Circle* (1971), a
trenchant analysis of ecological risks inflicted by use of inappropriate technologies.
Paul Ehrlich's *The Population Bomb* (1968) underscored the role of human population growth in resource use and environmental degradation. Rapid growth in the capacity of the nation's media to alert the public to such dangers heightened public
concern, and policy entrepreneurs like consumer activist Ralph Nader helped to tie
environmental quality to prominent health and safety issues.

The effect of these developments was a broadly based public demand for more
vigorous and comprehensive governmental action to protect valued natural resources and to prevent environmental degradation. New environmental organizations quickly arose and adopted the tactics of other 1960s social movements, using
well-publicized protests and university-based "teach-ins" to mobilize the public to

press for policy change. A variety of political reforms, including congressional redistricting and easier access by public interest groups to the courts and legislatures, and changes in the mass media, facilitated their success.

THE MODERN ENVIRONMENTAL MOVEMENT AND ITS POLICY ACHIEVEMENTS

The environmental movement of the late 1960s and 1970s represented one of those unusual periods in U.S. political history when the problem, policy, and politics streams converged, setting the stage for a dramatic shift in environmental policies. In an unprecedented fashion, a new environmental agenda rapidly emerged. It sought to expand and strengthen early conservation programs and to institute new public policies organized around the integrative and holistic concept of environmental quality. Environmental quality could bring together such otherwise distinct concerns as public health, pollution, natural resources, energy use, population growth, urbanization, consumer protection, and recreation (Caldwell 1970). That agenda drew from new studies of health and environmental risks and widespread dissatisfaction with the modest achievements of early federal air and water pollution control policies and equally limited state efforts (Davies and Davies 1975; Jones 1975). The legacy of this period includes the major federal environmental protection statutes, a host of important natural resource measures, and countless state and local initiatives that established environmental concerns firmly on the governmental agenda.

Contributing to the building of this new environmental policy agenda was an extraordinary group of policy entrepreneurs on Capitol Hill. They and their staffs had been patiently incubating these issues for years before public demand had crystallized and the media (and the White House) discovered the environment. These included such influential environmental lawmakers as Henry Jackson, Edmund Muskie, and Gaylord Nelson in the Senate, and Paul Rogers, John Saylor, Morris Udall, John Blatnik, and John Dingell, among many others, in the House. Even in early 1970, there was clear evidence of the problem stream changing quickly and influencing members of Congress who dealt with environmental policy. Indeed, the 92nd Congress (1971–1972) was the most productive in history for environmental protection and natural resources. It enacted measures on water pollution control, restrictions on ocean dumping, protection of sea mammals and coastal zones, regulation of pesticides, and noise control. Within a few years, other legislation followed on endangered species, drinking water quality, disposal of hazardous waste, control of toxic substances, and management of federal lands and forests. Table 3.3 lists some of the most important of these federal laws. With a rapid rise in public concern about the environment, extensive coverage in the media, and lobbying by new environmental groups, it is not surprising that the key statutes received strong bipartisan congressional support.

I turn to the key features of environmental and natural resource policies in Chapters 4 and 5. It is worth noting here, however, that the major federal environmental protection statutes adopted in the 1970s departed sharply from previous efforts even

Table 3.3 Major Federal Environmental Laws: 1964 to 1990

1964   Wilderness Act, PL 88-577

1968   Wild and Scenic Rivers Act, PL 90-542

1969   National Environmental Policy Act, PL 91-190

1970   Clean Air Act Amendments, PL 91-604

1972   Federal Water Pollution Control Act Amendments (Clean Water Act), PL 92-500

        Federal Environmental Pesticides Control Act of 1972 (amended the Federal Insecticide, Fungicide and Rodenticide Act (FIFRA) of 1947, PL 92-516

        Marine Protection, Research, and Sanctuaries Act of 1972, PL 92-532

        Marine Mammal Protection Act, PL 92-522

        Coastal Zone Management Act, PL 92-583

        Noise Control Act, PL 92-574

1973   Endangered Species Act, PL 93-205

1974   Safe Drinking Water Act, PL 93-523

1976   Resource Conservation and Recovery Act (RCRA), PL 94-580

        Toxic Substances Control Act, PL 94-469

        Federal Land Policy and Management Act, PL 94-579

        National Forest Management Act, PL 94-588

1977   Clean Air Act Amendments, PL 95-95

        Clean Water Act (CWA), PL 95-217

        Surface Mining Control and Reclamation Act, PL 95-87

1980   Comprehensive Environmental Response, Compensation, and Liability Act (Superfund), PL 96-510

1982   Nuclear Waste Policy Act of 1982, PL 97-425 (amended in 1987 by the Nuclear Waste Policy Amendments Act of 1987, PL 100-203)

1984   Hazardous and Solid Waste Amendments (RCRA amendments), PL 98-616

1986   Safe Drinking Water Act Amendments, PL 99-339

        Superfund Amendments and Reauthorization Act, PL 99-499

1987   Water Quality Act (CWA amendments), PL 100-4

1988   Ocean Dumping Act of 1988, PL 100-688

1990   Clean Air Act Amendments of 1990, PL 101-549

Note: A fuller list with a description of the key features of each act can be found in Vig and Kraft (1994), Appendix 1; natural resource policies are discussed in Chapter 5, and summarized in Table 5.1.

if members were not always certain of their probable effectiveness or cost. Environmental policy was "nationalized" by adopting federal standards for the regulation of environmental pollutants, action-forcing provisions to compel the use of particular technologies by specified deadlines, and tough sanctions for noncompliance. Congress would no longer tolerate the cumbersome and ineffective pollution control procedures used by state and local governments (especially evident in water pollution control). Nor was it prepared to allow unreasonable competition among the states created by variable environmental standards. To some critics, these distinctly nonincremental changes in environmental policy constituted "speculative augmentation." The new policies attempted to hasten technological developments and went beyond government's short-term capabilities, invited administrative delays, and imposed heavy burdens on industry and state and local governments ill-prepared to respond to the new demands (Jones 1975). However, the political attractiveness of these new policies in the warm glow of the early 1970s was obvious.

These policy developments coincided with the massive third wave of broader social regulation in the 1970s. It focused on health and safety issues as well as environmental quality, and it resulted in the formation of new federal agencies. These included the Occupational Safety and Health Administration (OSHA), the Consumer Product Safety Commission, and the EPA itself, which consolidated previously scattered government programs affecting the environment. The new social regulation differed from the old in important respects. It reflected a deep distrust of establishment organizations (especially the business community) and a determination to open the administrative process to public scrutiny. It led to a more activist or reformist orientation within the agencies, depending on the administration in power, and to extensive participation by new public interest groups such as consumer organizations and environmentalists (Harris and Milkis 1989; Berry 1977). It also led to costly regulations that often were not industry specific. By the 1980s, these very qualities contributed to efforts in the Reagan presidency to reverse policy advances of the 1970s, particularly in response to complaints of excessive and needlessly expensive regulation. Environmental regulations were prime targets of this criticism and retrenchment.

The new environmental movement that was so critical to bringing about these policy changes drew much of its political strength and moral force from the American public itself, which has continued to be one of the most important determinants of U.S. environmental politics. Anthony Downs (1972) captured the new power of environmental public opinion in comments about the "issue-attention cycle." Downs postulated that prior to the upsurge in public concern in the late 1960s, the nation was in a "pre-problem stage" on the environment. Such a stage exists when highly undesirable social conditions may exist but attract little public attention even if a few experts, public officials, or interest groups are alarmed by them. As a result of a series of catalytic events (for example, an oil-rig blowout off the coast of Santa Barbara, California, in 1969) and the publicity they receive, the public "suddenly becomes both aware of and alarmed about" the problem.

The dramatic rise in public concern over environmental quality in the late 1960s was confirmed by survey research as was its slow decline during the 1970s to a lower but still substantial level, and its striking resurrection in the 1980s during the

Reagan administration (Dunlap 1992; Mitchell 1984, 1990). Hazel Erskine (1972) described the initial rise as a "miracle of public opinion" because of the "unprecedented speed and urgency with which ecological issues have burst into American consciousness. Alarm about the environment sprang from nowhere to major proportions in a few short years" (120). Coverage of the environment by both print and electronic media followed a similar pattern, as did congressional agenda-setting activity such as hearings held on the subject (Baumgartner and Jones 1993). Despite a decline in the salience of environmental issues by the mid-1970s, the environment had become a core part of mainstream American values and was viewed almost universally by the public as a positive symbol with few negative images attached to it (Mitchell 1984, 1990; Dunlap 1989, 1992). It was as close to a consensual issue as one usually finds in U.S. politics.

Environmental Interest Groups

Both the older and the newer environmental lobbies found such a supportive public opinion to be an invaluable political resource in their lobbying campaigns in Washington and in state capitals. Many of the established conservation groups saw their membership soar between 1960 and 1972. The newer environmental organizations grew rapidly as well (Mitchell, Mertig, and Dunlap 1992). The trend continued through the 1970s, and then accelerated in the 1980s as environmental groups mounted highly successful membership recruitment campaigns in response to the antienvironmental agenda of the Reagan presidency. The groups' budgets and staffs grew in parallel with membership rolls (Baumgartner and Jones 1993, 189). The results can be seen in Table 3.4. One recent estimate puts the total membership of U.S. environmental groups at 14 million people, and their budgets at $600 million a year.

Despite their many shared values and political goals, environmentalists began to splinter into factions with conflicting styles and political strategies. The differences became so great that some analysts wondered whether they could be subsumed under the same label at all. For example, writing in 1991, Robert Mitchell noted that the unity of environmental groups was "tempered by a diversity of heritage, organizational structure, issue agendas, constituency, and tactics." They competed with each other, he said, "for the staples of their existence—publicity and funding" (Mitchell 1991, 83). Some of these divisions were noticeable even in the early 1970s.

At least three major categories of environmental groups should be distinguished: mainstream, greens, and grassroots. The mainstream organizations, such as the Sierra Club and the National Wildlife Federation, have evolved into highly professional Washington-based organizations that focus on public policy issues. In 1991 the National Wildlife Federation, for example, had 35 people in its Washington office assigned to tracking legislation. The more radical greens (including groups like Earth First! and the highly successful Greenpeace, with over 1.4 million members and a $33 million budget in 1991) tend to emphasize public education, direct action, and social change. The grassroots groups, which have sprung up in great numbers around the nation over the past decade (one 1993 estimate put the number at 12,000) and include those identified with the "environmental justice movement," deal

Table 3.4   **Membership and Budgets of Selected National Environmental Organizations, 1970–1990**

MEMBERSHIP

Organization	Year Founded	1960	1970	1975	1980	1985	1990	1990 Budget ($ million)
Sierra Club	1892	15,000	150,000	170,000	181,000	364,000	600,000	28
National Audubon Society	1905	32,000	105,000	275,000	400,000	550,000	575,000	40
National Parks and Conservation Association	1919	15,000	50,000	50,000	31,000	45,000	100,000	4
Wilderness Society	1935	10,000	44,000	50,000	35,000	150,000	350,000	14
National Wildlife Federation	1936	NA	3.1 mil.	3.7 mil.	4 mil.	4.5 mil.	5.8 mil.[a]	79
Environmental Defense Fund	1967	*	*	30,000	46,000	50,000	150,000	15
Natural Resources Defense Council	1970	*	*	35,000	42,000	50,000	125,000	13.5

Source: Christopher J. Bosso, "After the Movement: Environmental Activism in the 1990s," in *Environmental Policy in the 1990s,* 2nd ed., eds. Norman J. Vig and Michael E. Kraft (Washington: CQ Press, 1994), 36. Membership figures are notoriously hard to pin down. All figures reported here should be considered estimates and used only to illustrate growth over time. The 1990 estimates of organizational budgets are drawn from Margaret Kriz, "Shades of Green," *National Journal* (July 28, 1990): 1828.

[a] NWF membership figures include the large number of affiliated members. Without counting affiliates, NWF membership was approximately 975,000 in 1992.

* In 1970 neither the Environmental Defense Fund nor the Natural Resources Defense Council was a membership organization.

chiefly with local environmental issues such as threats from hazardous waste sites (Mitchell 1991; Bosso 1991).

Not all groups, however, fit easily into these categories. Some focus more on education, policy analysis, and scientific research than policy advocacy (for example, Resources for the Future, the Union of Concerned Scientists, the Worldwatch Institute, and the World Resources Institute). Others, such as the Nature Conservancy, concentrate on private land conservation and eschew direct political involvement.

The environmental movement in the 1990s is in the throes of change. Environmentalists face an array of new challenges and dilemmas that flow from their very success over the past two decades. The mainstream groups are now a political fixture in Washington and in many state capitals. Yet they and other environmentalists face difficult choices over organizational priorities and approaches. They are confronted by a newly energized opposition in the property rights and Wise Use movements, which reflect timber, mining, ranching, and other land development interests, particularly in the West (see Chapter 5). Environmentalists still face significant opposition by industries that traditionally have criticized strong environmental policies for their costs and other burdens they impose. The mainstream groups are losing ideological fervor to the greens and the emerging grassroots environmental groups. These and other disagreements over goals, strategies, and political styles—and competition for members and funds—continue to divide the environmental community (Gottlieb 1993). Over the past several years, most of the national groups have struggled to stem declining memberships and dwindling financial contributions. The Nature Conservancy has been an exception. In 1994 its membership reached 790,000, its highest level ever, and its fundraising rose by 15 percent to an all-time high.

Public Opinion on the Environment

There isn't much doubt about the public's continued concern about the environment and its broad support for public policy actions. Survey data indicate that the public believes that as a society we are subject to more risks in the 1990s than we were in the 1960s. Although people are living longer and generally have healthier and more comfortable lives today, the public is clearly nervous about newer risks that are different in both character and magnitude from those encountered in earlier decades. Surveys show public fear is especially great over radioactive wastes, toxic chemicals, hazardous wastes, and similar threats (Slovic 1987).

Public Concern and Support for Environmental Policy   The effects of these beliefs can be seen in the public's appraisal of environmental problems and support for environmental protection policies. When asked in 1988 whether they expected the outlook for a series of national problems to be better, about the same, or worse in another 15 years, the public placed the environment at the top of the list in problem severity (Mitchell 1990, 90). Many other surveys confirm that despite considerable societal and governmental efforts to improve the environment, the public believes environmental conditions from the local to the global level have worsened and that they will continue to decline in the future (Dunlap 1991). People believe such expected deterioration in the environment poses a direct threat to their health and well-being.

Even though people's environmental opinions are not usually very salient or central to their lives, these findings suggest there is a powerful new depth to environmental concern in the 1990s that was largely absent during the 1970s (Dunlap 1992). The public may be ready to respond to political leaders (and environmental activists) who learn how to appeal to their concerns, fears, and hopes for the future. A 1990 Roper survey found that 4 percent of the adult public (which is about 8 million people) claimed to "write letters to politicians expressing opinions on environmental issues" on a "regular basis," with 14 percent (27 million) doing so "from time to time" (Roper Organization 1990). The potential political influence of such an active public is enormous.

The same pattern of public commitment to environmental policy goals can be detected in a *New York Times*/CBS News survey conducted between 1981 and 1990. It found a nearly continual increase in the percentage of the public agreeing with the following strongly worded statement: "Protecting the environment is *so* important that requirements and standards cannot be too high, and continuing environmental improvements must be made *regardless* of cost." In September 1981, nine months into the Reagan administration, 45 percent of the public agreed with that statement and about the same number disagreed. By 1990, nearly 75 percent agreed and only one-fifth expressed disagreement (Bosso 1994). As Riley Dunlap (1987) has so carefully documented, the public was both aware of and generally disapproved of Reagan's environmental policies early in his administration. There was a significant shift upward in public concern for the environment during Reagan's presidency. The data, Dunlap concluded, indicate that Reagan "obviously failed to convince the public that environmental protection regulations need to be curtailed in order to achieve economic prosperity. Indeed, the administration's environmental policies and practices seem to have produced the opposite effect" (Dunlap 1987, 11).

Survey findings like these point to a "greening of America," even if the extent of this conversion may be questioned (Americans for the Environment 1989; Dunlap and Scarce 1991). As early as 1980, 62 percent of U.S. citizens surveyed in a government-sponsored study said they were "sympathetic" to the environmental movement or active within it; only 4 percent indicated they were unsympathetic (Council on Environmental Quality 1980). In an August 1991 *Wall Street Journal*/NBC poll, eight in ten Americans said they regarded themselves as "environmentalists." One-half of the respondents said they were "strong" ones.

One recent survey tried to stratify the public in a way that may be more revealing of public attitudes and behavior. It found only about 11 percent of the public could be described as "True-Blue Greens" or committed and active environmentalists. Another 11 percent were termed "Greenback Greens." They shared strong environmental values with the True-Blues, and were prepared to pay more for "green" products. Yet they were much more reluctant to sacrifice their time or convenience to protect the environment. About 52 percent of the public in this survey were characterized as indifferent to environmental concerns and inactive on the issues (Roper Organization 1990).

International opinion surveys conducted in 1992 a few months before the Earth Summit confirmed that strong support for environmental values is by no means confined to the industrialized nations. The Gallup Health of the Planet survey covered two dozen nations, from low-income to high-income, and it found there was "little difference in reported levels of environmental concern between people of poor, less economically developed nations and those of the richer, highly industrialized nations." Majorities in 21 of the 24 nations surveyed reported either a "great deal" or a "fair amount" of concern (Dunlap, Gallup, and Gallup 1993, 11). More remarkably, residents of poor nations were only slightly more willing than those in rich nations to accept environmental degradation in exchange for economic growth.

How Deep Is Public Concern for the Environment?   Such survey findings notwithstanding, the public's personal behavior does not always match its stated opinions on the environment, as the 1990 Roper survey indicated. As a journalist once put it, "tree talk" is cheap, but it creates a "lip-service gap" when the public is unwilling to make the necessary personal sacrifices to clean up the environment.

This is especially so in the 1990s, when environmental policy increasingly will call for changes in personal behavior (e.g., in the way we commute to work, recycle waste products, and use energy and water) as opposed to regulation of industry.

Moreover, the public often lacks knowledge of technical issues and is not fully able to judge the various warnings of environmental hazards and "crises" reported in the press and pressed on them by environmental organizations. A 1993 survey sponsored by Defenders of Wildlife found that 73 percent of U.S. adults were unfamiliar with the issue of loss of biodiversity. Even after it was explained to them, few associated the problem with loss of habitat. Many other examples of such limited environmental knowledge could be cited.

In addition to these complications in interpreting public opinion, the public's views are inconsistent. People favor greater environmental regulation and yet want "less government." Or they support energy conservation in theory, but are unwilling to pay even modest sums in increased gasoline taxes to further it. The public's general preference for environmental protection also may weaken when people face real and intense local or regional conflicts in which environmental measures are believed to adversely affect employment or economic well-being.

For all these reasons, some question the depth of public commitment to environmental policy goals seemingly revealed in survey research (Dunlap 1991). It is not clear how readily people would change their behavior if given the necessary information, encouragement, and incentives. The success of many local recycling programs indicates considerable potential may exist. It is equally evident that environmental organizations have not yet succeeded in building the firm basis of public support they will need to advance their agenda over the next several decades. They will have to do a lot better at public education and in expanding their efforts to reach constituencies they have largely ignored to date, including urban residents, minorities, and middle-class citizens whose dedication to environmental values remains fairly shallow.

Environmental Issues in Election Campaigns   One way to increase public support and reach new constituencies is to promote environmental issues in election campaigns. Unfortunately, environmental issues only rarely have been a decisive factor in election campaigns, even though they have long been prominent in selected contests in New Jersey, California, and several other states. In an effort to alter that pattern, environmental groups such as the Sierra Club have created separate political action committees (PACs) to contribute funds to green candidates. In 1992 the Sierra Club's PAC spent about $600,000, with club-endorsed candidates winning over 60 percent of Senate races and 76 percent of those for the House. For over two decades, the League of Conservation Voters (LCV) has served as a national center for fundraising and electoral support for environmentally oriented candidates (chiefly those running for Congress). In 1992 the League spent $600,000 (the same as the Sierra Club), and it hoped for a $4 million budget for the 1994 congressional races.

Despite these attempts to influence elections, the relatively minor role played by environmental issues in most election campaigns sends a message to elected (and other) public officials that they have much more leeway in decision making than environmental opinion polls appear to indicate. While members of Congress generally have been reluctant to vote against popular environmental bills, the legislative process offers innumerable opportunities to satisfy the public while meeting the objections of industry or other critics through obscure statutory provisions. The trick for environmental leaders and policy entrepreneurs is to focus on pertinent issues that strike close to home and mobilize the electorate around their genuine concerns. Environmentalists try to use this electoral lever in their lobbying. Carl Pope, executive director of the Sierra Club, spoke to the point in 1994 as environmentalists faced a powerful political backlash and new barriers in renewing major laws:

> Much as we might like to be able to move a positive agenda by inspiring people, providing good ideas, working with enlightened and sympathetic people in the Administration and Congress . . . the ground rules remain that in order to actually move environmental legislation, those who oppose it must feel pain. And they must feel pain where they care about it, which is back home (Cushman 1994g, B7).

POLITICAL REACTION IN THE 1980s: THE REAGAN AND BUSH ADMINISTRATIONS

If the decade of the 1970s reflected the nation's initial commitment to environmental policy, the 1980s indicated at least some ambivalence about how far to go in pursuit of policy goals, and at what price. Such concerns had surfaced in the Nixon and Ford administrations, and again in the late 1970s in the Carter administration, in part in association with the Arab oil embargo of 1973 and further energy and economic shocks of the decade (Whitaker 1976; Vig 1994). Industry had complained about the financial burdens of environmental regulation, and some labor unions joined in the chorus when jobs appeared to be threatened.

These criticisms reached full force in Ronald Reagan's presidency (1981 to 1989). Reagan and his advisers brought with them a dramatically different view of environmental issues and their relationship to the economy than had prevailed, with bipartisan support, for the previous decade. Virtually all environmental policies were to be reevaluated, and reversed or weakened, as part of the president's larger political agenda. That agenda included reducing the scope of government regulation, cutting back on the role of the federal government, shifting responsibilities where possible to the states, and relying more on the private sector.

Set largely by the extreme right wing of the Republican party and western ranchers and others associated with the Sagebrush Rebellion (see Chapter 5), Reagan's agenda was the most radical in half a century (Kraft 1984). The major problem he faced was that Congress and the American public continued to favor the programs he was so anxious to curtail. Reagan tried as much as he could to bypass Congress in pushing his agenda through an administrative strategy of deregulation, defunding of regulatory agencies, and appointment of high-level personnel more in tune with his own conservative ideology. That strategy was only moderately effective and soon backfired.

Although initially successful in gaining congressional support for deep budgetary cuts that he justified largely on economic grounds, Reagan soon found his deregulatory actions sharply criticized and blocked by Congress. EPA administrator Anne Burford was cited for contempt of Congress over her handling of hazardous waste programs and her refusal, on presidential orders, to deliver subpoenaed documents to Congress. She was forced to resign after less than two years in office. Reagan replaced her in March 1983 with William D. Ruckelshaus, who had served as the first EPA administrator between 1970 and 1973. Interior Secretary James G. Watt, the administration's controversial point person on environmental policy, lasted only a little longer before resigning in October 1983. He was replaced by William P. Clark, who, like Ruckelshaus, was more moderate and sought to repair some of the political damage Watt caused (Vig and Kraft 1984).

Congress went on to renew and fortify every major environmental statute that came up for renewal in the 1980s. These included the Resource Conservation and Recovery Act (RCRA) in 1984, the Superfund, and the Safe Drinking Water Act in 1986, and the Clean Water Act in 1987. The RCRA in particular was markedly strengthened in direct response to the political turmoil at the EPA and congressional distrust of the agency. By 1990, even the Clean Air Act, the nation's most important environmental law, was expanded and strengthened with the support of Reagan's vice president and successor, George Bush. The reason for this broad support for the legislation lay in good measure with public opinion and the capacity of environmental leaders to capitalize on the public's favorable stance toward environmental policy and its distrust of both industry and government. Early in 1990, for example, public opinion surveys indicated that over 70 percent of the U.S. public favored making the Clean Air Act stricter, over 20 percent favored keeping it about the same, and only 2 percent wanted it less strict (National Journal 1990).

Budget cuts, personnel shifts, administrative changes, and lax enforcement of the laws during the 1980s took their toll on the EPA and other governmental agencies and on achievement of environmental goals (Vig and Kraft 1984; Vig 1994). Adjusted for inflation, for example, the EPA's operating budget by fiscal 1990 was just about the same as it had been in 1975 before many of the new environmental laws were in force. The EPA's staff at the end of Reagan's term in January 1989 was only slightly larger than it had been at the beginning of the decade despite increasing responsibilities given to the agency by Congress. Overall federal spending on environmental and natural resource programs, adjusted for inflation, fell by about 10 percent from 1980 to 1989. It had fallen by over 30 percent by 1986, but Congress began increasing budgets again in the late 1980s (Vig and Kraft 1994).

In retrospect, the remarkable turmoil over environmental policy under Ronald Reagan, and under George Bush as well, can be seen as a short-term retrenchment in a striking evolutionary advance toward ever stronger environmental policies in the United States and globally. As discussed above, public support not only continued in the 1980s, it strengthened considerably. Environmental groups saw their memberships grow to new heights and media attention was mostly favorable to environmentalists.

The message was not lost on George Bush's political advisers. During his run for the presidency in 1988, Bush broke openly with the Reagan environmental agenda.

In a remarkable speech on the shore of Lake Erie in late August 1988, he promised to be "a Republican president in the Teddy Roosevelt tradition. A conservationist. An *environmentalist*" (Holusha 1988).[8]

Bush's record on the environment was decidedly mixed, though it clearly was different from Reagan's in many respects. Bush's chief handicap was his tendency to defer to his conservative economic and political advisers, particularly chief of staff John H. Sununu, budget director Richard Darman, and economic adviser Michael Boskin. Like Reagan's White House advisers, they viewed environmental policy with great suspicion. Bush's Interior, Energy, and Agriculture departments largely continued to favor greater development of natural resources even while EPA head William K. Reilly pressed for stronger environmental protection efforts.

The Bush administration suffered from deep internal divisions over the direction of environmental policy that contributed to the president's inability to live up to his initial promise to depart sharply from Reagan's agenda. From energy policies that neglected conservation to rejection of a U.S. leadership role on international population and environmental policies and his stubborn opposition to a strong climate change treaty at the 1992 Earth Summit, Bush's presidency ultimately was a great disappointment to the environmental community. He ended his presidency openly critical of "environmental extremists," and vice-presidential candidate Al Gore in particular. Bush didn't convince the American public. A Gallup poll in June 1992 indicated that the public disapproved of his handling of the environment by a two to one margin (Vig 1994).

CONCLUSION

In at least one respect, the Reagan and Bush administrations left an important legacy for environmental policy that will be addressed throughout the rest of the 1990s. Environmental policy has to be judged by many of the same standards applied to other public policies. Good intentions are not enough. We expect government to offer demonstrable evidence of success. Where programs are ineffective or highly inefficient, alternatives need to be examined. If environmental standards are to be set at high levels, some indication of positive benefits needs to be offered to justify the costs and burdens imposed on society. All these considerations lie at the heart of current efforts to appraise and reform environmental policy. In the next two chapters, I turn to the goals and chosen instruments of U.S. environmental protection policy and natural resource and energy policy, and to debate over issues of this kind. Chapter 6 takes an even more direct tack on trying to assess the effectiveness and impacts of two decades of U.S. environmental policy.

The record of policy evolution reviewed in this chapter points to an important paradox of U.S. environmental policy and politics. Policy achievements of the 1960s and 1970s withstood a severe challenge during the 1980s in large part because the American public strongly favored prevention of further environmental degradation and protection from environmental threats. In this sense, the U.S. political system has proved to be fairly responsive to public demands despite its institutional fragmentation and other barriers to majority rule. The paradox is that government may be too responsive to apparent public *demands* and insufficiently attentive to long-term public *needs*. In response to public opinion, Congress has mandated that the EPA and other agencies take on many more tasks than they are equipped to do. Environmental interest groups have effectively lobbied for adoption of those policies and their strengthening over time. The problem is that elected officials have failed to specify priorities among the dozens of measures adopted that would aid agencies in their implementation of them. They also have neglected to evaluate those policies carefully and to ensure that they can achieve the goals set.

As discussed in Chapter 1, this tendency to reflect public fears about the environment and to overload and underfund administrative agencies translates into significant inefficiencies in U.S. environmental policy. We spend a lot of money and time dealing with problems that pose little risk to public and ecological health and not nearly enough on other problems, such as climate change, loss of biological diversity, and indoor air quality, of far greater consequence. So democracy works, sort of. An important question is how U.S. government and politics can be improved so that policies in the future better serve the public's long-term needs and ensure a sustainable and just environment for future generations both at home and in the rest of the world that is inevitably affected by U.S. policy choices.

ENDNOTES

1. The policy process model is not without its critics. Some argue that it should be replaced by a more accurate and genuinely causal model of policy activities that lends itself to empirical testing and to incorporation of a broader variety of policy actors and behavior (Sabatier and Jenkins-Smith 1993). I disagree with much of this critique. The policy cycle model can be highly useful for describing the diversified players in the policy game and for alerting us to pertinent actions that contribute to our understanding of environmental politics and policy. We should not, however, treat the stages in the model as anything more than helpful constructs.
2. For an extended analysis of environmental policy entrepreneurship, particularly with regard to the National Environmental Policy Act and environmental impact assessment, see Wandesforde-Smith (1989).
3. For a detailed and sophisticated review of U.S. government institutions and environmental policy, see the edited collection by Lester (1995). Leading environmental policy scholars assess the characteristics and performance of Congress, the bureaucracy, the courts, and state government as well as public opinion, interest groups, and political parties. Chapters in Vig and Kraft (1994) also offer appraisals of Congress, the presidency, the courts, the EPA, and federalism.
4. The Fifth Amendment states, among other things, that no person shall "be deprived of life, liberty, or property, without due process of law; nor shall private property be taken for public use, without just compensation."
5. The EPA's budget is further disadvantaged because within the House it is governed by the Veterans Affairs, Housing and Urban Development, and Independent Agencies Subcommittee, a body with a highly diversified jurisdiction.
6. In addition to Shabecoff's history of the environmental movement, many recent volumes on U.S. conservation history provide a full account of what can only be summarized briefly here. See Fox (1985), Lacey (1989), Nash (1990), and Hays (1959, 1987). The

volume edited by Lacey is especially useful for students of environmental policy and politics. It offers extensive coverage of social and economic changes leading to the development of modern environmental policy. Robert Mitchell's chapter recounts the birth of, and important changes in, the environmental movement. Other chapters survey the development of policies on parks and wilderness, public lands, wildlife and endangered species, and toxic substances, among others.

7. These public lands are managed chiefly by the National Park Service (responsible for about 25 million acres), the U.S. Forest Service (191 million acres), and the Bureau of Land Management (450 million acres). The balance of land outside of Alaska consists of desert, rangeland, and mountain land.

8. President Bush's statement of August 31, 1988 is reprinted in an article he wrote for *Sierra*, "Promises to Keep," in the November–December issue, p. 116.

CHAPTER FOUR

Environmental Protection Policy: Controlling Pollution

The American public has made clear its desire for clean air, clean water, and a healthy environment free of toxic substances and hazardous wastes. It has been less explicit about its eagerness to pay higher taxes and product costs or to suffer the bureaucratic interventions that go with the territory. Congress has responded to public opinion by enacting, and over time strengthening, seven major environmental protection, or pollution control, policies whose purpose has been to achieve those broad goals: the Clean Air Act, Clean Water Act, Safe Drinking Water Act, Toxic Substances Control Act, Federal Insecticide, Fungicide, and Rodenticide Act, Resource Conservation and Recovery Act, and Comprehensive Environmental Response, Compensation, and Liability Act or Superfund. The first six look primarily to the future. The EPA develops regulations that affect the current and future generation, transportation, use, and disposal of pollutants that pose a risk to public health or the environment. The last is largely, though not exclusively, remedial. It aims to repair damage from careless disposal of hazardous chemicals in the past. Superfund also affects present and future pollution control efforts.

Separately and collectively, these acts mandate an exceedingly wide array of regulatory actions that touch virtually every industrial and commercial enterprise in the nation and increasingly will directly affect the lives of ordinary citizens. Routine implementation decisions as well as the periodic renewal of the acts by Congress invariably involve contentious debates over the extent of the risks faced, the appropriate standards to protect public health and the environment, the policy instruments used to achieve those standards, and the degree to which environmental and health benefits should be weighed against the costs of compliance and other social and economic values.

As shown in Chapter 2, these policies have produced significant and well-documented gains, particularly in urban air quality and in control of point sources of water pollution. It is equally true, however, that the policies have fallen short of expectations, and in some cases distressingly so, as actions on toxic chemicals and hazardous wastes illustrate. Existing policy barely touches some major risks, such as indoor air quality, and others, such as surface water, groundwater, and drinking water quality, have proved far more difficult to control than originally expected. As desirable as they are, the achievements are not cheap. Cumulative expenditures between 1972 and 1992 ran in excess of $1 trillion (U.S. General Accounting Office 1992).

The nation remains committed to the broad policy goals of controlling pollution and minimizing public health risks. Yet implementation costs and the increasing intrusiveness of environmental regulation over the next several decades will test its resolve. Tight budgets at all levels of government and growing impatience with ineffective and inefficient public policies are creating new demands that programs either produce demonstrable success or be reconfigured in some way (Schneider 1993a, d).

THE CONTOURS OF ENVIRONMENTAL PROTECTION POLICY

This chapter reviews the configuration of current environmental protection policy to provide some perspective on these developments and on contemporary policy debates. Each policy has different goals, uses distinctive means to achieve them, and sets out its own standards for balancing environmental quality and other social goals. Yet they may also be grouped together for present purposes. All focus on environmental protection or pollution control, and all seek primarily to protect public health even where ecological objectives are included as well. All also rely on national environmental quality standards and regulatory mechanisms, and all are implemented primarily by the federal EPA, in cooperation with the states. I pay particular attention to the institutional and other actors who shape policy decisions, the political and administrative processes that affect the outcomes, and competing approaches and solutions. I save some of the larger questions of how well these programs are working for Chapter 6.[1]

The Clean Air Act

The Clean Air Act (CAA) is one of the most comprehensive and complex statutes ever approved by the federal government. It is the premier example of contemporary environmental regulation. There should be no surprise, therefore, that since the early 1970s the act has been a frequent target of critics even while environmentalists and public health specialists have sought to fortify it. The battles have continued into the 1990s. Congress approved the 1990 amendments with President George Bush's active support. Yet his own White House Council on Competitiveness (discussed below) tried repeatedly to weaken EPA regulations for implementing the act in an effort to reduce its economic impact on industry and local governments (Bryner 1993, 173–178; U.S. General Accounting Office 1993b).

Federal air pollution control policy dates back to the original and modest Clean Air Act of 1963, which had provided for federal support for air pollution research as well as assistance to the states for developing their own pollution control agencies.

Prior to the 1963 act, federal action was limited to a small research program in the Public Health Service authorized in a 1955 act. The 1970 amendments to the Clean Air Act followed several incremental adjustments in federal policy, including the 1965 Motor Vehicle Air Pollution Control Act and a 1965 amendment to the CAA that began the federal program for setting emission standards for new motor vehicles. A 1967 Air Quality Act provided funds to the states to plan for air pollution control, required them to establish air quality control regions (geographic areas with shared air quality problems), and directed the federal government to study health effects of air pollution to assist the states in their control strategies.

Congress adopted the radically different 1970 CAA both in response to sharply increased public concern about the environment and because it saw little progress under the previous statutes in cleaning the air. The states as well as the federal government had been slow in responding to worsening air quality problems, states were reluctant to use the powers they were given, and the automobile industry displayed little commitment to pollution control. With near unanimity, Congress approved a vastly stronger act despite intense industry pressure to weaken it and significant doubts about both available technology and governmental capacity to implement the law (Jones 1975; Bryner 1993).

The new policy mandated National Ambient Air Quality Standards (NAAQSs) that would be uniform across the country and set by the EPA, with enforcement shared by the federal and state governments. The primary standards were to protect human health and secondary standards where necessary would protect buildings, forests, water, crops, and similar nonhealth values. The EPA was to set the NAAQSs at levels that would "provide an adequate margin of safety" to protect the public from "any known or anticipated adverse effects" associated with six major, or criteria, pollutants: sulfur dioxide, nitrogen dioxide, lead, ozone, carbon monoxide, and particulate matter. The standard for lead was added in 1977. A standard for hydrocarbons that the EPA issued under the 1970 act was eliminated in 1978 as unnecessary.

The act also required that an "ample margin of safety" be set for toxic or hazardous air pollutants such as arsenic, chromium, hydrogen chloride, zinc, pesticides, and radioactive substances. Congress assumed that a safe level of air pollution existed and standards could be set accordingly for air toxics as well as the criteria pollutants. The act has been interpreted as requiring the setting of the crucially important NAAQSs without regard to the costs of attainment (Portney 1990b). If control technologies were not available to meet these standards, Congress expected them to be developed by fixed deadlines. The goal was to reduce air pollution to acceptable levels.

The 1970 CAA and subsequent amendments also set national emissions standards for mobile sources of air pollution: cars, trucks, and buses. Congress explicitly called for a 90 percent reduction in hydrocarbon and carbon monoxide emissions from the levels of 1970, to be achieved by the 1975 model year, and a 90 percent reduction in the level of nitrogen oxides by the 1976 model year. Yet it gave the EPA the authority to waive the deadlines—which it did several times. Similarly, the act set tough emission standards for stationary sources such as refineries, chemical companies, and other industrial facilities. New sources of pollution were to be held to New Source Performance Standards to be set industry by industry and enforced by the states. The standards were to be based on use of state-of-the-art control technologies (best available technology, or BAT), with at least some recognition that economic costs and energy use should be taken into account. Existing sources (e.g., older industrial facilities) were held to lower standards set by the states.

To deal with existing sources of air pollution, each state was to prepare a State Implementation Plan (SIP) that would detail how it would meet EPA standards and guidelines. States have primary responsibility for implementing those sections of the act dealing with stationary sources; the EPA retains authority for mobile sources. The 1970 act called for all areas of the nation to be in compliance with the national air standards by 1975, a date later extended repeatedly. The EPA could reject a SIP if could not be expected to bring the state into compliance, and the agency could impose sanctions on those states, such as banning construction of large new sources of air pollution, including power plants and refineries, or cutting off highly valued federal highway and sewer funds.[2]

In the 1977 amendments to the Clean Air Act, Congress backtracked in some respects from its early uncompromising position—for example, on dates by which compliance was to be achieved. Yet it left the major goals of the law intact and even strengthened the act on requirements for nonattainment areas and provisions for prevention of significant deterioration (PSD) in those areas already cleaner than the national standards. Congress established three classes of clean areas. In Class I (national parks and similar areas), air quality would be protected against any deterioration. In Class II, the act specified the amount of additional pollution that would be permitted. Finally, in Class III, air pollution was allowed to continue until it reached the level set by national standards. Congress further provided for protection and enhancement of visibility in national parks and wilderness areas which were being affected by haze and smog. A new and controversial provision in the 1977 amendments called for the use of "scrubbers" to remove sulfur dioxide emission from new fossil-fuel burning power plants whether those plants used low- or high-sulfur coal. The action was widely understood to be an effort to protect the high-sulfur coal industry (Ackerman and Hassler 1981).

The nation made substantial progress in meeting policy goals during the 1970s, and the results were evident in cleaner air in most American cities. Nevertheless, unhealthy levels of air pollution continued and little progress was being made on control of toxic air pollutants called for in the 1970 act. Newer problems, such as acid rain and the contribution of greenhouse gas emissions to climate change, demanded attention. At the same time, criticism mounted over the EPA's implementation of the act, particularly its inconsistent, inflexible, and costly regulations and inadequate guidelines for state action. Technical obstacles to meeting the act's goals and deadlines also became apparent. Conflicts over reformulation of the act were so great that Congress was unable to fashion a compromise acceptable to all parties until 1990.

The 1990 amendments to the CAA further extended the act's reach to control of the precursors of acid rain (sulfur dioxide and nitrogen oxides) emitted primarily by coal-burning power plants and to the CFCs that damage the ozone layer. Among the most innovative provisions in Title IV of the 1990 amendments was the use of an emissions trading program for reducing sulfur dioxide emissions. Title II of the amendments called for further reductions (of 35 to 60 percent) in automobile tailpipe emissions between 1994 and 1996, and mandated development of cleaner fuels for use in selected areas of the nation. Oxygen-containing additives are to be used in fuels in communities with high levels of carbon monoxide (chiefly in wintertime when carbon monoxide is a problem). Reformulated, or cleaner, gasoline is to be used in cities with severe ozone problems. The reformulated fuels require more refining and will cost slightly more, perhaps 4 to 6 cents a gallon. They will burn more thoroughly and evaporate more slowly, and will contain much lower concentrations of toxic compounds like toluene and benzene. These fuels will also have oxygen-containing additives.

Because residents of many metropolitan areas are still forced to breathe polluted air, Title I of the act set out an elaborate and exacting multitiered plan intended to bring all urban areas into compliance with national air quality standards in three to 20 years, depending on the severity of their air pollution. The amendments invited private lawsuits to compel compliance with those deadlines.

Unhappy with the EPA's dismal progress in regulating hazardous air pollutants, Congress departed sharply from the 1970 act. It required the agency to set emission limits for all major industrial sources of hazardous or toxic air pollutants (e.g., chemical companies, refineries, and steel plants). These rules are intended to reduce emissions by as much as 90 percent by the year 2003 through technology-based standards. Title III lists 189 specific toxic chemicals that are to be regulated as hazardous air pollutants. Within eight years of setting emission limits for industrial operations, the agency is required to set new health-based standards for those chemicals determined to be carcinogens representing a risk of one cancer case in one million exposed individuals.

In Title V of the act, Congress established a new permitting program to facilitate enforcement of the act. Major stationary sources of air pollution are required to have EPA-issued operating permits that specify allowable emissions and control measures that must be used (Bryner 1993).

As attention shifted from Congress to the intricacies of the EPA's rule making and state implementation efforts, controversy continued through 1994, particularly during the Bush administration (see below). The broad scope and demanding requirements of the 1990 act guarantee that conflict is not going to fade away anytime soon.[3]

The Clean Water Act

The Federal Water Pollution Control Act Amendments of 1972, now known as the Clean Water Act (CWA), is the major policy regulating surface water quality. As was the case with the Clean Air Act, the CWA dramatically altered the original and very limited Water Pollution Control Act (1948), which emphasized research, investigations, and surveys of water problems.[4] Under the 1948 act, there was no federal authority to establish water quality standards or restrict discharge of pollutants. Amendments adopted in 1956 also failed to establish any meaningful control over discharge of pollutants. The 1965 Water Quality Act went much further by requiring states to establish water quality standards for interstate bodies of water and implementation plans to achieve those standards, and by providing for federal oversight of the process. Yet the 1965 law was widely viewed as administratively and politically unworkable and ineffective, in part because of significant variation among the states in their economic resources, bureaucratic expertise, and degree of commitment to water quality goals.

The 1972 CWA was intended to correct these deficiencies by setting a national policy for water pollution control. It established deadlines for elimination of the discharge of pollutants into navigable waters by 1985 and stated that all waters were to be "fishable and swimmable" by 1983. It also encouraged technological innovation and areawide planning for attainment of water quality (Freeman 1990). The 1972 act was itself revised and strengthened in 1977 and 1987 without fundamentally altering the goals or means to achieve them. However, Congress postponed several deadlines for compliance and established new provisions for toxic water pollutants, the discharge of which in "toxic amounts" was to be prohibited. In 1994 and 1995 Congress was deeply immersed in the latest round of revisions and renewal of the act.

The goals of the 1972 act may have been admirable, but they proved to be wildly unrealistic. The tasks were of a staggering magnitude, industry actively opposed the act's objectives and frequently challenged the EPA in court, technology proved to be costly, and planning for control of nonpoint sources (those with no specific point of origin) was inadequately funded and difficult to establish. Thus deadlines were postponed and achievement of goals suffered. For example, in 1992, the International Joint Commission reported that the United States had yet to eliminate completely the discharge of any persistent toxic chemical. The achievement of fishable and swimmable waters is supposed to mean, as stated in the 1977 act, that the water quality provides for "the protection and propagation of a balanced population of shellfish, fish, and wildlife" as well as recreation in and on the water. As noted in Chapter 2, the EPA reported in 1994 that about 40 percent of rivers and lakes and one-third of estuaries assessed failed to meet those standards. Nevertheless, as the first sentence of the Clean Water Act states, it aspires to "restore and maintain the chemical, physical, and biological integrity of the nation's waters." Elaborate efforts are underway to move toward those challenging objectives and signs of progress are evident.

Much like the Clean Air Act, the Clean Water Act gives the states primary responsibility for implementation as long as they follow federal standards and guidelines. Dischargers into navigable waterways must meet water quality standards and effluent limits, and operate under a permit that specifies the terms of allowable discharge and control technologies to be used. The EPA has granted authority to most states to issue those permits, which operate under the National Pollutant Discharge Elimination System (NPDES). The NPDES applies to municipal wastewater treatment facilities as well as to industry. Compliance with the permits is determined by self-reported discharge data and on-site inspections by state personnel. Unfortunately, noncompliance has been a significant problem (Russell 1990). The states also establish Water Quality Criteria (WQC) that define the maximum concentrations of pollutants that are allowable in surface waters. In theory, this means that the WQC concentrations would be set at levels posing no threat to individual organisms, populations, species, communities, and ecosystems (including humans). States do this using EPA guidelines that take into account the uses of a given body of water as defined by the states (e.g., fishing and boating—and waste disposal). The purpose is to prevent degradation that interferes with the designated uses. States may consider benefits and costs in establishing water quality standards.

Effluent limitations specify how much a given discharger is allowed to emit into the water and the specific treatment technologies that must be used to stay within such limits. The EPA has defined such effluent limitations for different categories of industry based on available treatment technologies. Economists and industry leaders have long objected that the requirements are irrational: "To put it simply, standards are set on the basis of what can be done with available technology, rather than what should be done to achieve ambient water quality standards, to balance benefits and costs, or to satisfy any other criteria" (Freeman 1990, 106). Critics also object to what they consider to be vague statutory language such as "best practical," "best available," and "reasonable costs," which grant enormous discretion to administrative agencies to interpret complex and varied scientific, engineering, and economic information. In response, defenders of the CWA argue that use of approaches that call for best available technology (BAT) is a major reason why the nation has achieved the degree of progress it has under the act.

A politically attractive feature of the Clean Water Act was federal funding to assist local communities to build municipal wastewater treatment facilities. The federal government initially assumed 75 percent of the capital costs, and for several years in the 1970s it subsidized such construction to the tune of $7 billion per year. The percentage of federal assistance was reduced in the 1980s, and in recent years, the amounts have been closer to $2.5 billion per year. The generous subsidies are unlikely to continue much longer. They are slated for replacement by a revolving state loan program that will provide seed money for plant construction.

The discharge of toxic chemicals has proved to be much more difficult to regulate than conventional pollutants. Although bodies of water may assimilate a certain amount of biologically degradable waste products, this is not the case with toxic chemicals, which often accumulate in toxic hot spots in river and lake sediments. By the mid-1990s, consensus was emerging that the best course was to identify and end the use of the most toxic, persistent, and bioaccumulated pollutants. For example, this position was a cornerstone of the EPA's Great Lakes Five-Year Strategy. The agency's proposed Great Lakes Water Quality Initiative called for the "virtual elimination" of discharges of persistent toxic substances throughout the Great Lakes basin.

Unlike conventional or "point" sources such as industry discharge pipes, nonpoint sources of water pollution such as agricultural runoff have proved to be exceptionally difficult to manage. The 1987 CWA amendments required the states to develop an EPA-approved plan for control of nonpoint sources such as urban storm-water runoff, cropland erosion, and runoff from construction sites, woodlands, pastures, and feed lots. Before implementation of the CWA, these sources constituted between 57 percent and 98 percent of total discharges of phosphorus, nitrogen, suspended solids, and biological oxygen demand in the nation's surface waters (Freeman 1990, 109). Even by 1994, they remained responsible for about two-thirds of stream pollution. So far, most states have chosen voluntary approaches using "best management practices" to deal with nonpoint sources.

Eventually, as water quality controls are ratchetted up, new technologies are mandated, and industries alter their production processes through pollution prevention programs, water quality should improve. A key question is how much the development of such control technologies should be pushed. The very threat of higher standards and thus higher costs has persuaded many companies to concentrate on pollution prevention. With preventive approaches, a toxic chemical is eliminated through substitution of a less toxic or nontoxic one, or the industrial process itself is changed—for example, through wastewater recycling and filtration.

The way in which Congress will change the Clean Water Act remains uncertain. During 1994, a half-dozen coalitions were fighting one another over the act's provisions. Business groups organized as the Water Quality Task Force (chemicals, oil, and durable goods), the Clean Water Industry Coalition (agriculture and manufactures), and the Clean Water Working Group (agricultural chemicals). Environmentalists marched under the banner of the Clean Water Network (organized by the Natural Resources Defense Council and including some 450 other groups). The Clean Water Network released a report entitled the *National Agenda for Clean Water* that called for sweeping reforms in the act. Environmentalists also mounted a forceful grassroots lobbying effort to try to win the day. State and local governments concerned about the high costs of water policies formed their own coalition, which included the National Association of Counties, National Governors' Association, National Conference of State Legislators, National League of Cities, and U.S. Conference of Mayors (Carney 1994). Congress failed to reach agreement on renewal of the CWA in 1994 and it tried again in 1995.

The Safe Drinking Water Act

The 1974 Safe Drinking Water Act (SDWA) was designed to ensure the quality and safety of drinking water by specifying minimum public health standards for public water supplies. The act authorized the EPA to set National Primary Drinking Water Standards for chemical and microbiological contaminants for tap water. The act also required regular monitoring of water supplies to ensure that pollutants stayed below safe levels.

The EPA made slow progress in setting standards. Only 22 standards for 18 substances had been set by the mid-1980s. In 1986 a Congress frustrated with both the EPA's pace of implementation and insufficient action by state and local governments strengthened the act. Congress required the EPA to determine maximum contaminant levels for 83 specific chemicals by 1989 and set quality standards for them, set standards for another 25 contaminants by 1991, and 25 more every three years. Congress was highly prescriptive in detailing what contaminants would be regulated, how they would be treated, and the timetable for action. The standards are to be based on a contaminant's potential for causing illness and the financial capacity of medium- to large-sized water systems to foot the bill for the purification technology.

States have the primary responsibility for enforcing those standards for over 50,000 public water systems in the United States, most of which serve small communities with fewer than 10,000 people. Water systems were to use the best available technology to remove contaminants and monitor for the presence of a host of chemicals. Even the EPA acknowledges that the states receive less than half the funds needed to comply. The problems are especially severe for the thousands of small water systems that can ill afford the cost of expensive new water treatment technologies. The EPA estimates that through the year 2000, small systems will incur costs of almost $3 billion to comply with all regulations, and $20 billion more to repair, replace, and expand their water system infrastructures (U.S. General Accounting Office 1994a). For those reasons, the nation's governors and mayors pressed Congress in 1994 to ease regulatory red tape by focusing on tests for contamination and monitoring of only those chemicals posing the greatest risk to human health. The drinking water law symbolized a broader complaint about so-called "unfunded federal mandates" that require states and localities to spend their scarce local funds on environmental programs over which they have no say. The SDWA also reflected concern over the high marginal costs of further improvements in environmental quality after the gains of the past 20 years (Kriz 1994b).

With the SDWA up for reauthorization in 1994, the Safe Drinking Water Act Coalition, representing a dozen organizations of state and local officials, lobbied aggressively for reducing the regulatory burden on states. It sought the use of less expensive and slightly less effective technology, less strict water quality standards where public health would not be endangered, and a new four-year $3.6 billion federal revolving loan fund to defray the cost for smaller water systems. Environmentalists were as determined to keep stringent public health protections in place (Kriz 1994b).

Groundwater contamination is of special concern for municipalities that rely on well water and for the nation's rural residents. Aquifers may be contaminated by improper disposal of hazardous wastes and leaking underground storage tanks, and from agricultural runoff (for example, nitrates and pesticides), among other sources. Groundwater historically has been governed chiefly by state and local governments. Yet it is affected by the Safe Drinking Water Act, the Clean Water Act and many other federal statutes, with no consistent standards or coordination of enforcement.

Even though compromise legislation to renew the SDWA in 1994 garnered the support of a broad array of interest groups, House and Senate negotiators failed to reach agreement on the measure. The newly proposed revolving state fund to help small communities pay for water treatment facilities was lost in the process. As the case with other environmental policies up for renewal in 1994, the 104th Congress will return to consideration of the act in 1995.

The Resource Conservation and Recovery Act

Although the federal government had dealt earlier with solid waste in the 1965 Solid Waste Disposal Act (SWDA), by the 1970s, concern was shifting to hazardous waste. In 1976 Congress enacted the Resource Conservation and Recovery Act (RCRA) as amendments to the SWDA and to the 1970 Resource Recovery Act. RCRA was to regulate existing hazardous waste disposal practices as well as to promote the conservation and recovery of resources through comprehensive management of solid waste. Congress addressed the problem of abandoned hazardous waste sites in the 1980 Superfund legislation discussed below.

RCRA required the EPA to develop criteria for safe disposal of solid waste and the Commerce Department to promote waste recovery technologies and waste conservation. The EPA was to develop a "cradle-to-grave" system of regulation that would monitor and control the production, storage, transportation, and disposal of wastes considered hazardous, and it was to determine the appropriate technology for disposal of wastes.

The act delegated to the EPA most of the tasks of identifying and characterizing such wastes (the agency counts over 500 chemical compounds and mixtures) and determining whether they are hazardous. That judgment is to be based on explicit measures of toxicity, ignitability, corrosivity, and chemical reactivity. If a waste is positive by any one of these indicators, or is a "listed" hazardous waste, it is governed by RCRA's regulations (Dower 1990).

RCRA also established a paper trail for keeping track of the generation and transportation of hazardous wastes that was intended to eliminate so-called "midnight" or illegal dumping. Eventually, the EPA developed a national manifest system for that purpose, but since most hazardous waste never leaves its site of generation, the manifest system governs only a small portion of the total. More importantly, RCRA called for the EPA to set standards for the treatment, storage, and disposal of hazardous wastes that are "necessary to protect human health and the environment." As is the case for the air and water acts, the EPA over time delegated authority to most states to implement RCRA.

Initially, the EPA was exceedingly slow in implementing RCRA, in part because of the unexpected complexity of the tasks, lack of sufficient data, and staff and budget constraints. In addition, there were frequent battles among environmentalists, EPA technical staff, the chemical industry, and the White House (especially in the Reagan administration) over the stringency of the regulations (Cohen 1984). The EPA took four years to issue the first major regulations, and two more years to issue final technical or performance standards for incinerators, landfills, and surface storage tanks that had to be met for licensed or permitted facilities. In the meantime, public concern had escalated due to publicity over "horror stories" involving disposal of hazardous waste at Love Canal, New York, and soil contaminated with dioxin-tainted waste oil at Times Beach, Missouri. EPA relations with Capitol Hill were severely strained in the early 1980s during Anne Burford's tumultuous reign as administrator, and the Reagan administration ultimately had no legislative proposal for renewal of RCRA that would allow the EPA to put forth its own vision of a workable policy.

As a result of these developments, Congress grew profoundly distrustful of EPA's "slow and timid implementation of existing law," and it sharply limited administrative discretion in its 1984 rewrite of RCRA, officially called the Hazardous and Solid Waste Amendments Act (HSWA). The 1984 RCRA amendments rank among the most detailed and restrictive of environmental measures ever enacted, with 76 statutory deadlines, eight of them with "hammer" provisions that were to take effect if the EPA failed to act in time (Halley 1994). Congress uses "hammer" language in environmental and other acts to impose a legislative regulation if an agency fails to adopt its own regulations by the stated deadline. Such provisions are intended to force agency compliance with the law, but they may interfere with sound policy implementation, particularly when agency budgets are tight and much technical uncertainty surrounds the problems being addressed.

The 1984 act sought to phase out disposal of most hazardous wastes in landfills by establishing demanding standards of safety; expand control to cover additional sources and wastes (particularly from small sources previously exempt); extend RCRA regulation to underground storage tanks (USTs) holding petroleum, pesticides, solvents, and gasoline; and move much more quickly toward program goals by setting out a highly specific timetable for mandated actions. The effect of all this was to dramatically drive up the cost of hazardous waste disposal. While economists question the economic logic of these provisions (Dower 1990), Congress helped to bring about an outcome long favored by environmentalists: the internalization of environmental and health costs of improper disposal of wastes. If disposal of wastes is extraordinarily expensive, there is a powerful incentive to produce less of them, thus leading (eventually) to source reduction, recycling, and new treatment technologies, which the 1984 amendments ranked as far more desirable than land disposal. At least that's the idea.

Congress struggled in the early 1990s with reauthorization of RCRA without resolving continuing conflicts. Through 1992, the Bush administration opposed revamping of the act, saying it was unnecessary. Members of Congress, however, heard warnings of impending solid waste crises because most of the nation's remaining landfills will close over the next 15 years. Environmentalists pressed for higher levels of waste reduction and waste recovery through recycling and tighter controls on waste incineration. Local governments worried about what to do with incinerator ash that may be classified as hazardous under a 1994 Supreme Court decision, and some states fought the solid waste industry over efforts to restrict interstate transport of solid wastes. As was often the case over the past decade, environmental gridlock prevailed as each side fought for its preferred solutions and no consensus emerged on a comprehensive revision of RCRA.

The Toxic Substances Control Act

After five years of development and debate, Congress enacted the Toxic Substances Control Act (TSCA) in 1976. The EPA was given comprehensive authority to identify, evaluate, and regulate risks associated with the full life cycle of commercial chemicals, both those already in commerce as well as new ones in preparation. The TSCA aspired to develop adequate data on the effect of chemical substances on health and the environment and to regulate those chemicals posing an "unreasonable risk of injury to health or the environment" without unduly burdening industry and impeding technological innovation.

The EPA was to produce an inventory of chemicals in commercial production, and it was given authority to require testing by industry where data are insufficient and the chemical may present an unacceptable risk. However, exercise of that authority was made difficult and time consuming. An Interagency Testing Committee, with representatives from eight federal agencies, was established to recommend candidates and priorities for chemical testing. Where data are adequate, the EPA may regulate the manufacture, processing, use, distribution, or disposal of the chemical.

Options range from banning the chemical to labeling requirements, again with demanding, formal rule-making procedures required (Shapiro 1990).

Congress also granted to the EPA the authority to screen new chemicals. The agency must be notified 90 days before manufacture of a new chemical substance, when the manufacturer must supply any available test data to the agency in a Premanufacturing Notice. If the EPA determines the chemical may pose an unreasonable risk to health or the environment, it may ban or limit manufacture until further information is provided. The requirements here are more easily met than for existing chemicals, and the EPA can act more quickly.

Although the meaning of "unreasonable risk" is not formally defined in the act, Congress clearly intended some kind of balancing of the risks and the benefits to society of the chemicals in question (Shapiro 1990, 211). The TSCA was modified by amendments in 1986, the Asbestos Hazard Emergency Response Act, and in 1992 by the Residential Lead-Based Paint Hazard Reduction Act. The former required the EPA to develop strategies for inspecting schools for asbestos-containing material and controlling the risk appropriately. The latter called for a variety of actions to reduce public exposure to lead from paint. These include inspection and abatement of lead hazards in low-income housing, disclosure of the risks of lead-based paint prior to the sale of homes built before 1978, and development by the EPA of a training and certification program for lead abatement contractors.

Like the other major acts, the implementation of TSCA has not gone smoothly. The EPA encountered resistance from industry in getting the necessary information, had difficulty recruiting sufficient trained personnel for the regulatory tasks, and made very slow progress in achieving TSCA's objectives. The EPA's job was made more difficult than it otherwise might have been by forcing the agency to prove that a chemical was unsafe or posed an unreasonable risk. As a result of these stipulations in the law and the other constraints the agency faced, only a handful of chemicals have been banned under TSCA.

The Federal Insecticide, Fungicide, and Rodenticide Act

Federal regulation of pesticides is much older than laws dealing with other chemical risks. It dates back to a 1910 Insecticide Act designed to protect consumers from fraudulent products. In 1947 Congress enacted the Federal Insecticide, Fungicide, and Rodenticide Act (FIFRA), authorizing a registration and labeling program, and it gave authority for its implementation to the Department of Agriculture. Concern focused chiefly on the efficacy of pesticides as agricultural chemicals. By the 1960s, following Rachel Carson's *Silent Spring,* public attention shifted to environmental consequences of pesticide use. Congress amended FIFRA in 1964, 1972, and 1978, establishing the present regulatory framework. Jurisdiction over the act was given to the EPA in 1970.

FIFRA requires that pesticides used commercially within the United States be registered by the EPA. It sets as a criterion for registration that the pesticide not pose "any unreasonable risk to man or the environment, taking into account the economic, social, and environmental costs and benefits of the use" (Shapiro 1990). The law is less stringent than other environmental statutes of the 1970s, and critics assert that the government has given greater weight to economic arguments than to the impacts on public health. The EPA is required to balance costs and benefits, with the burden of proof of harm placed on the government if the agency attempts to cancel or suspend registration of an existing pesticide. For a new pesticide, the burden lies on the manufacturer to demonstrate safety. Procedures under the law are cumbersome, however, making regulatory action difficult. These statutory provisions reflect the still considerable power of the pesticide lobby despite the many gains environmental, health, and farm worker groups have made against the long-influential agricultural subgovernment (Bosso 1987).

The EPA is also required under the Food, Drug, and Cosmetic Act to establish maximum permissible concentrations of pesticides in or on both raw agricultural products (for example, fresh fruits and vegetables) and processed foods. Those standards are then enforced by the Food and Drug Administration (FDA) and the Department of Agriculture. The EPA is governed in part by the so-called "Delaney clause" in the act that bans any food additive shown to cause cancer in laboratory animals. Of the roughly 400 kinds of pesticides used on food crops, more than 70, including the most commonly used, cause cancer in laboratory animals. In a long awaited 1993 report, the National Academy of Sciences indicated that children may be at special risk from exposure to trace levels of pesticides from foods as well as from lawn care products and household insect sprays. Representative Henry Waxman (D-California), chair of the Health and the Environment subcommittee in the House, termed the report a "wake-up call to all of us" about the "unnecessary risk from pesticides in food" (Michaelis 1993).

Much recent controversy over FIFRA concerns whether and in what way to modify the Delaney clause. To bring the law into conformity with practice and with contemporary views of the relatively minor risks to public health posed by minute pesticide residues in food, the Clinton administration proposed in 1993 a more realistic "negligible risk" standard. It would allow no more than one additional case of cancer for every one million people. The food and chemical industries support such an easing of the Delaney policy, but many environmental groups oppose the action.

Because of its origins in the agricultural policy community, FIFRA differs in many respects from other environmental laws. Indeed, environmentalists have referred to it as an "anachronistic statute" that is "riddled with loopholes and industry-oriented provisions." Although reauthorized in 1988 after years of political controversy and legislative stalemate, the act reflected only modest changes, leading critics to dub it "FIFRA Lite." In recent years, the Campaign for Pesticide Reform, a coalition of environmental, health, consumer, and labor groups, had sought an array of more far-reaching changes in the act.

By 1994, national environmentalists joined with a diversity of grassroots organizations to form a new coalition called the National Campaign for Pesticide Policy Reform. These groups have proposed speeding up the renewal or cancellation of pesticide registration, strengthening the law to help control contamination of groundwater, limiting pesticide residues in food, protecting farm workers from exposure to harmful chemicals, and improving public access to health and safety information. Industry groups, organized as the National Agricultural Chemicals Association (which in 1994 changed its name to the American Crop Protection Association), have protested that such changes would cost millions of dollars, yield no appreciable benefits, and jeopardize trade secrets.

The impasse continued in 1994 after the Clinton administration submitted its own FIFRA renewal proposals to Congress without strong support from either of the two sides. In addition to modifying the standard of allowable risk in processed and raw foods, the administration's measures included increased reliance on integrated pest management and other alternatives to pesticide use—in part through educational programs aimed at farmers. The administration also backed review of all pesticide registrations every 15 years, prohibition on export for any pesticide banned in the United States for health reasons, easing of the standards for scientific and legal proof for withdrawing a pesticide from the market if suspected of posing an unacceptable risk to health or the environment, and elimination of economic consequences as a criterion for determining the degree of hazard presented by a pesticide. The effect of these changes would be to eliminate the use of dozens of the leading pesticides within three to seven years. Farmers, pesticide makers, and food processors opposed the changes on the grounds that the U.S. food supply is already among the safest in the world.[5] A wide gulf exists between opposing interests, and debate over FIFRA is likely to continue for years to come.

The Comprehensive Environmental Response, Compensation, and Liability Act

Congress enacted the Comprehensive Environmental Response, Compensation, and Liability Act (CERCLA), better known as Superfund, in 1980 and revised it in 1986 with the Superfund Amendments and Reauthorization Act (SARA). The act is a partner to RCRA. Whereas RCRA deals with current hazardous waste generation and disposal, the Superfund is directed at the thousands of abandoned and uncontrolled hazardous waste sites. Little was known about the number, location, and risks associated with these sites, and existing law was thought to be insufficient to deal with the problem. With Superfund, Congress gave the EPA responsibility to "respond" to the problem by identifying, assessing, and cleaning up those sites. The EPA could use, where necessary, a special revolving fund of $1.6 billion, most of which was to be financed by a tax on manufacturers of petrochemical feedstocks and other organic chemicals and crude oil importers. The act put responsibility for the cleanup and financial liability on those who disposed of hazardous wastes at the site, a "polluter pays" policy.

Unhappy with the pace of cleanup and the Reagan administration's lax implementation of Superfund, in 1986 with SARA, Congress authorized an additional $8.5 billion for the fund and mandated stringent cleanup standards using the best available technologies. SARA also established an entirely new Title III in the act, stimulated by the 1984 chemical plant accident in Bhopal, India, that killed over 2000 people and injured hundreds of thousands. Within three months of the Bhopal accident, bills in Congress merged the right-to-know concept with the Superfund reauthorization legislation.[6] Title III, also called the Emergency Planning and Community Right-to-Know Act (EPCRA), provided for public release of information about chemicals made by, stored in, and released by local businesses (published each year as the Toxics Release Inventory). It also required the creation of state and local committees to plan for emergency chemical releases (Hadden 1989).

Superfund Provisions and Controversies   Superfund gives the EPA authority to identify the parties responsible for inactive or abandoned hazardous waste sites and to force cleanup. It may also clean up sites itself, and seek restitution from the responsible parties. Such actions are governed by complicated "strict, joint, and several liability" provisions of the act that can be especially burdensome on minor contributing parties. The parties responsible may be sued as a group or individually for all of the cleanup costs even if they are not at fault. The act's retroactive liability provision also holds companies liable for wastes disposed of legally prior to 1980.

In addition, Superfund requires that sites be identified and ranked according to their priority for cleanup. The EPA and the states nominate sites for a National Priority List (NPL), and the EPA uses a hazard ranking system to measure the severity of the risks at each site. The rankings do not reflect actual human or environmental exposures, however, but rather potential health and environmental risks as judged by EPA project officers and technical consultants (Dower 1990; Mazmanian and Morell 1992, 31). Only sites listed on the NPL qualify for long-term cleanup under Superfund and for use of the federal dollars associated with the program.

The EPA has evaluated close to 24,000 nonfederal sites (i.e., not on property owned by the federal government) for possible inclusion on the NPL, and it has begun work on another 9000. Together they represent 94 percent of the nonfederal sites brought to the agency's attention (U.S. Congress 1994). Although critics point to the small number of NPL sites that have been fully cleaned up, defenders of the program note that long-term remediation has begun at over one-half of the NPL sites and that all have been stabilized to some degree to eliminate immediate threats (De Saillan 1993). In 1994 cleanup at about 50 to 70 sites was being completed each year, with about 50 other sites added to the NPL list. As of early 1994, almost 1300 sites were on the list.

The full process—from identification and preliminary assessment of a site to hazard ranking, listing, remedial design, and remediation itself—is complex and time consuming. It can take as long as 15 years (some state cleanups take only two or three), which helps explain the slow pace of actual cleanup. Groundwater cleanup can be especially difficult and costly. Other reasons for the limited progress have included extensive litigation among potentially responsible parties and insurance companies and the enormous difficulty in locating hazardous waste treatment and disposal sites due to community opposition (Mazmanian and Morell 1992; Hird 1994; U.S. Office of Technology Assessment 1988, 1989).

Controversy over the Superfund program has centered on the degree of cleanup needed for any given site (and hence the cost). Under present policy, cleanup costs have averaged about $30 million per NPL site, but with wide variability of costs depending on the nature of the risks at the site and the cleanup standard used. Total costs for all expected Superfund sites could be as high as $350 billion if standards are stringent or as "low" as $90 billion under standards less demanding than current policy (Russell, Colglazier, and Tonn 1992). Those costs would be paid for chiefly by the responsible parties, not the federal government.

Proposed Reforms   After consulting for months with business and environmental interests through a National Commission on the Superfund (also known as the Keystone Commission), the Clinton administration in early 1994 proposed reauthorization legislation to revamp the highly criticized program (Cushman 1994a). Included were some easing of cleanup standards depending on the intended future use of the site, a consideration ignored in the original policy, and an alteration in the liability sections of the law that critics (including affected businesses and insurance companies) have labeled "unworkable." It must be said, however, that some assessments of CERCLA have found these liability provisions to be one of the strengths of the law. They have produced "new political, financial, legal, and managerial incentives" that are profoundly restructuring the way the nation handles hazardous wastes and public access to information about them (Bartlett 1994, 178).

Industry also had complained that returning all sites to pristine or "greenfield" conditions was unnecessarily burdensome if they are intended for further industrial use. Business groups also saw little point in being forced to clean up trace amounts of chemicals that posed little or no measurable health risk. Environmentalists wanted uniform standards for all sites nationwide and generally opposed the flexibility that industry sought in these standards. However, they were more favorably inclined if local communities would have a significant say in cleanup decisions. Grassroots groups associated with the environmental justice movement actively sought early citizen participation in these decisions to ensure that minority communities were ranked high in priority for cleanup actions (Kriz 1994a). The revised law will likely include a national standard on the maximum public health risk that is tolerable after cleanup for the most common 100 chemicals at such sites. However, such a standard would fall short of returning all sites to pristine conditions. Further negotiations in mid-1994 between environmentalists and the business community, assisted by the Clinton White House, seemed to resolve most of the remaining differences over Superfund reauthorization (Kriz 1994d).[7]

The key policy actors on Superfund were in greater agreement on how liability for cleanup costs could be allocated among responsible parties, in part through use of an independent arbitrator. By mid-1994, there was also consensus among the major insurance companies to establish a new $8 billion fund based on a tax on their industry to help pay for site cleanup, which was expected to greatly reduce litigation. Litigation and negotiation costs have consumed a significant amount of the money spent on Superfund sites (about 24 percent according to the Congressional Budget Office), although disputes exist over exactly how much. Despite the promising outlook for reform of Superfund in 1994, Congress failed to enact the new legislation and it returned to the task in 1995.

THE INSTITUTIONAL CONTEXT OF POLICY IMPLEMENTATION

As should be clear at this point, these seven statutes present the EPA with an astonishingly large and bewildering array of administrative tasks that are essential to meet congressional mandates for environmental protection. It would be truly remarkable if the agency could pull it all off, and especially if it could keep Congress and the multitude of constituency groups happy with the results. That the EPA shares responsibility for implementing these statutes, and about a dozen others, with the states is a mixed blessing that creates supervisory headaches even while it relieves some of the routine burdens of administration.

Critics of regulatory policy often lump all agencies together. Yet they are a highly varied lot, and their individual characteristics must be considered to understand why they operate as they do. The EPA's success or failure in policy implementation is affected by most of the usual factors shaping administrative decision making and some that are distinctive to the agency (Bryner 1987). Some of these are largely beyond the control of agency officials, such as the intractability of environmental problems, changing economic conditions and technology, budgetary resources, statutory authority (e.g., sanctions and incentives for inducing compliance), and political judgments made by Congress and the White House. Others are influenced to at least some extent by agency behavior. The administrative and leadership skills of agency officials significantly affect staff recruitment and expertise, internal organization and priorities, cooperation elicited from other federal agencies, and the political support received from the White House and Congress. Through their policy choices and public outreach efforts, EPA officials also can shape the public's attitudes toward environmental issues and the agency's legitimacy and competency in the eyes of important policy actors such as the environmental community, business, and state and local governments (Landy, Roberts, and Thomas 1990).

The EPA's Organization, Budget, and Staff

The EPA's organizational structure, budgetary resources, and staff characteristics are especially important for policy implementation. President Richard Nixon created the agency by executive order on December 2, 1970, following submission of a reorganization plan to Congress. The order transferred most (though not all) of the existing federal environmental programs to the EPA, which was established as an independent executive agency. Its administrator and other top officials are nominated by the president and confirmed by the Senate. Unlike environmental ministries in other western democracies, the EPA does not yet enjoy cabinet rank, although the administrator is the only head of a regulatory agency reporting directly to the president. Proposals to convert the agency into a Department of the Environment, with cabinet status, have languished in Congress for years due to controversies over the direction of environmental policy. The physical location of the agency in a remote corner of Southwest Washington in two converted apartment buildings has symbolized the EPA's uncertain status in the universe of federal agencies. After years of promises of a new headquarters in downtown Washington, the EPA staff were scheduled to begin moving into new quarters on Pennsylvania Avenue in 1996.

As shown in Figure 4.1, the EPA's organization reflects its media-specific responsibilities, with separate program offices for air and radiation, water, pesticides and toxic substances, and solid waste and emergency response. Each has operated independently despite a widely acknowledged imperative of instituting integrated or cross-media approaches to pollution control. In July 1994, however, EPA Adminis-

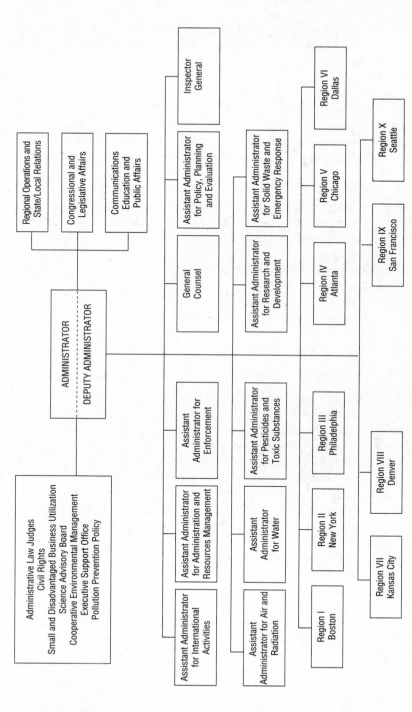

Figure 4.1 **Organization Chart of the U.S. Environmental Protection Agency**

trator Carol M. Browner announced a Common Sense Initiative to begin shifting away from a pollutant-by-pollutant approach in the separate media. The agency intends to review pollution across air, water, and land within each major industry sector to help coordinate rules and regulations.

As is the case with many federal agencies, much of the EPA's routine implementation work takes place in its ten regional offices. Two-thirds of the agency's staff are employed in those offices, where they work closely with state governments. Rule making and policy development are the responsibility of EPA headquarters staff in Washington.

In its first full year in 1971, the EPA had a staff of about 7000, a total budget of $3.3 billion, and an operating budget (the funds used to implement its programs) of $512 million (Portney 1990a, 10). It has grown considerably larger over time, as have its responsibilities, and it is by far the largest federal regulatory agency. By 1994, the staff had increased to nearly 18,000, the overall budget to about $6.5 billion, and the operating funds to about $2.7 billion. According to the GAO, however, the fiscal 1993 EPA operating budget in constant dollars was about the same as it was in 1979 (U.S. General Accounting Office 1993d). For much of the past 20 years, the EPA has been deprived of the resources needed to implement environmental protection policies successfully. The EPA's current budget would be in far worse shape had it not been raised substantially since 1990 under Presidents Bush and Clinton. For fiscal 1995, the Clinton administration recommended a boost of 8 percent in the overall budget and 13 percent in operating funds.

The EPA tends to recruit a staff strongly committed to its mission, but with a diversity in professional training and problem-solving orientations that breeds conflict over implementation strategies (Landy, Roberts, and Thomas 1990). The staff consists predominantly of scientists and engineers, who are assisted by a multitude of lawyers, economists, and policy analysts. With the notable exception of Anne Burford in the early 1980s, the agency has attracted talented and respected administrators: William Ruckelshaus (twice), Russell Train, Douglas Costle, Lee Thomas, and William Reilly. After her first year as administrator, Clinton-appointee Carol Browner received far fewer accolades and was overshadowed as administration environmental spokesperson by Interior Secretary Bruce Babbitt. By the end of 1994, however, lawmakers on the Hill praised Browner for her persistent efforts to reform environmental protection policy despite the low priority President Clinton appeared to give to the issues.

As might be expected for an agency dealing with demanding technical problems, the EPA has developed a reputation for expertise even while the quality of its science is disparaged. In recent years, the agency has had 12 research labs and 28 facilities that provide technical support for its regulatory efforts. It also has made extensive use of outside scientists. Its Science Advisory Board (SAB), created by Congress in 1978, has been, according to a former director, the "principal ongoing, institutionalized mechanism by which the scientific community interacts with all levels of EPA on many issues" (Yosie 1993, 1476–77). The SAB advises the agency as requested through a series of specialized committees, conducts an annual review of the scientific adequacy of EPA's research and development (R&D) work, and reviews the technical quality of proposed criteria documents, standards, and regulations.

Nevertheless, numerous studies have found the agency's scientific work to be inadequately staffed and insufficient. They also have faulted the EPA for lacking a strategy for long-term environmental research essential to its mission (Carnegie Commission 1992, 1993b). In 1994 the agency began a major reorganization of its science laboratories in an effort to improve research capabilities. The organization of EPA science is not the only problem. Although federal environmental R&D spending totaled $4.5 billion in fiscal 1992, the EPA's share of the pie was relatively small at about $350 million (Gramp 1994). Altogether, EPA's in-house research funds run to about $500 million a year, still a small amount in the context of a total federal R&D budget of some $73 billion, and where the Defense Department R&D budget stands at about $40 billion a year. The agency has fewer scientists (1800) than it had in 1971, though it relies much more now on outside scientists under contract to the agency. The EPA's most vociferous critics have expressed their displeasure over the past decade by calling for more "science-based" (by which they often mean less stringent) environmental regulation. Yet many of them voiced little complaint when the EPA's R&D budget suffered steep cuts in the 1980s.

Despite its fairly large overall staff, the EPA has been forced to rely heavily on outside scientific consultants and contractors for some of its most visible activities, such as implementation of the Superfund program. By many accounts, it has done a poor job of supervising the contractors, who handle about one-third of the agency's work (U.S. General Accounting Office 1992). The EPA also is forced to work closely with other federal agencies such as the Food and Drug Administration, Occupational Safety and Health Administration, Department of Energy, and National Oceanic and Atmospheric Administration. Dependence on them adds to the administrative complexity and uncertainty of implementation.

Working with the States

Perhaps most importantly, the EPA has delegated to the 50 states the authority to implement most of the major environmental protection programs. They operate under the general supervision of the agency's Washington headquarters and especially under the regional offices, which keep in close contact with state officials and other environmental policy actors at the state and local level. Among the most important of those local actors are citizen activists and environmental watchdog groups who monitor and review key implementation actions such as the issuance of air and water pollution permits and new regulations. In some of the more environmentally progressive states, a kind of symbiotic relationship has developed between the state agency and citizen groups. Agency employees need the citizen groups to bring sufficient political pressure on state governments to push for more aggressive implementation than might otherwise occur. Citizen groups in turn may receive grants for public education and pollution prevention initiatives and serve on local, regional, and state stakeholder advisory committees on which agencies increasingly rely. In the less environmentally progressive states, business and industry groups may have the dominant influence in such decision-making processes.

States differ significantly in their programs, rules, and regulations, and in their capabilities for effective implementation. Yet most pass EPA muster and are granted primacy for program operations. The EPA supports those activities through a variety of program grants that rarely are sufficient to meet state and local needs (U.S. Congress 1988). In early 1995 both houses of Congress voted overwhelmingly to require that new legislative proposals (including major revisions of existing laws) imposing costs of more than $50 million annually on state and local governments be accompanied by a cost estimate to be prepared by the Congressional Budget Office. Sponsors also must specify how the programs would be financed. The law erects new procedural barriers to enacting statutes which are likely to lead to unfunded federal mandates.

In addition to a shortage of federal grant funds, the states have had other complaints about this complex system of environmental federalism. Chief among them have been late issuance of regulations and inflexibility in the detailed federal grant conditions and mandatory policy "guidelines" which waste state resources, stifle initiative, and add unnecessary costs. Other frequently cited problems include delayed program grants, excessive paperwork that diverted limited state staff, unclear regulations and guidelines, and friction between the EPA and state staff related to program deadlines and priorities (Kraft, Clary, and Tobin 1988; Hamilton 1990; Tobin 1992).

Some states are better able than others to make up for the shortfall in federal grants. Some of them also have demonstrated more interest in and capacity for innovative approaches to environmental efforts (Lowry 1992). By one ranking, the top ten states in environmental policy innovation are (in order) California, Oregon, New Jersey, Connecticut, Maine, Wisconsin, Minnesota, New York, Massachusetts, and Rhode Island (Lester 1994).[8]

Political Support and Opposition

All of the factors discussed here affect the EPA's ability to perform its many statutory tasks. Political support for the agency, and opposition to it, are equally important. The EPA has benefitted from strong support for environmental protection goals in Congress, among the public at large, and among the well-organized environmental community. That support has allowed it to fend off critics and to prevent weakening of its statutory authority. Members of Congress consider the EPA, like other federal regulatory agencies, to be a creature of their own making, and they keep close watch over its operation. They also defend it from presidents critical of its mission, as they did during the 1980s, although they frequently engage in EPA "bureau bashing" themselves to score political points about burdensome regulation (Rosenbaum 1994). Congress may grant or withhold administrative discretion to the EPA depending on the prevailing trust of the agency's leadership and confidence in the president. As it demonstrated with its reauthorization of RCRA in 1984 and with the CAA of 1990, Congress may choose to specify environmental standards and deadlines in great detail to compel the EPA to do what it wants (Halley 1994; Bryner 1993).

The critics of EPA decision making have not been without considerable political clout. Industry and state and local governments concerned about the high cost of environmental protection have found receptive ears in Congress in recent years among both Democrats and Republicans. Since the EPA's creation, presidents have taken a keen interest in its activities, and each has devised some mechanism that allows close White House supervision of the agency's regulatory program (Shanley 1992; Vig 1994; Harris and Milkis 1989).

The EPA has at best a mixed record of success, which reflects both the scope and complexity of the environmental problems it addresses and its organizational features. After a period of institutional growth and development in the 1970s, the agency suffered severe budget and personnel cuts in the early to mid-1980s, which took a toll on staff morale. Since then, the EPA has struggled to redefine its mission, improve its capabilities for risk assessment and environmental management, and to cope with the enormously difficult jobs given it by the U.S. Congress. Many students of environmental policy are convinced it still has a long way to go (Rosenbaum 1994; Landy, Roberts, and Thomas 1990). If its responsibilities continue to greatly exceed the resources provided to it and if the agency is unable to institute needed reforms, it will continue to disappoint the environmental community as well as the regulated parties. A review of environmental rule making and enforcement illustrates some of the problems that the agency faces.

SETTING ENVIRONMENTAL STANDARDS

At the heart of environmental policy implementation is the setting and enforcing of environmental standards. The process involves a number of distinctive scientific, analytic, and political tasks that cut across the different policy areas. These include the determination of overall environmental policy goals and objectives; the setting of environmental quality criteria, quality standards, and emissions standards; and enforcement of the standards through incentives and sanctions provided in the statutes.

Environmental Goals and Objectives

Environmental policy goals and objectives are set by elected public officials, especially Congress, and reflect their conception of how environmental quality is to be reconciled with other social and economic values. Goals and objectives historically have been general rather than specific ("fishable and swimmable waters") and ambitious, considering available technologies and resources. They emphasize important symbolic values and emotionally charged images (clean air, clean water, public health, safe drinking water) to maintain public support even when statutory language embodies inevitable compromises over competing values. Congress grants authority to agencies like the EPA to fill in the details by providing for administrative discretion in developing rules and regulations (Bryner 1987). By the 1980s, as we saw above, Congress grew impatient with the EPA and set highly specific goals and objectives and tighter deadlines, often with "hammer" clauses meant to greatly reduce agency discretion. However, Congress is rarely clear in establishing priorities among the environmental policy goals it sets for the EPA and other agencies. It seems content to say "do it all" even when it fails to appropriate the funds necessary for the job.

The EPA has no statutory charter or organic law that defines its mission and explicitly sets out its policy objectives. That omission is a major liability and is attributable to the agency's creation by executive order. The consequence is that the EPA administers ten major environmental statutes adopted at different times by different congressional committees for diverse reasons. It is not a comprehensive environmental agency, and it would not be even if it were made into a cabinet department. That's because significant environmental and natural resource responsibilities have been given to other agencies scattered across the executive branch. These deficiencies make the EPA more vulnerable than other agencies to short-term forces such as changes in public opinion, shifting legislative majorities, and presidential agendas. It also complicates the job of setting agency priorities and allocating scarce resources absent congressional specifications about which program should take precedence.

Environmental Quality Criteria

Environmental quality criteria lie chiefly in the scientific realm. They spell out what kinds of pollutants are associated with adverse health or environmental effects. In making such determinations, the EPA and other agencies such as the FDA and OSHA draw from experimental and epidemiological studies and modeling exercises from within and outside of government. Setting environmental criteria requires some kind of risk assessment to answer key questions. What is the relationship between pollution and health? For example, how do fine particulates affect the lungs? How do specific pollutants affect the functioning of ecosystems? At what level of contamination for toxic chemicals such as PCBs or dioxin can we detect either human health or ecosystem effects? Studies of such relationships eventually allow government agencies to set environmental protection criteria.

Health Risk Assessment   Because these determinations set the stage for regulatory action (or inaction), risk assessment has been at the center of disputes over environmental protection policy for well over a decade. There is no shortage of scientific controversies over the most appropriate models, assumptions, and measurements even in areas where there has been extensive experience with these methods, such as testing potential carcinogens (Andrews 1994; Rushefsky 1986). There is no single "right way" to conduct assessments that can completely resolve these arguments. This means that users of such assessments have to be alert to their limitations and rely on open debate over risk studies, and on adversarial processes, to detect and compensate for their weaknesses.

Health risk assessments normally involve a series of analytic procedures: (1) a *hazard identification* to determine whether a substance may be a health hazard; (2) a *dose-response assessment* to determine the relationship between the magnitude of exposure and the degree of a health effect; (3) an *exposure assessment* to determine how many people (and which segments of the population) may be exposed to what amounts, from what sources, how frequently, and for how long; and (4) a *risk characterization* that uses the information collected to describe the overall health risk, such as the increased chances of developing cancer (National Research Council 1983).

None of these activities is easy, and a good deal of uncertainty characterizes the whole process. At each step there are myriad scientific and policy judgments required to determine which methods to use and how to infer human health risk from limited epidemiological evidence or data drawn from animal exposures. The data gathering itself is complex, time-consuming, and expensive, and there is always a potential for misinterpretation. A major EPA study of dioxin released in 1994 illustrated all those problems and will likely be the subject of considerable debate over the next several years (Lee 1994).[9] Assessments also tend to concentrate on individual pollutants and on cancer risks. Synergistic effects of multiple pollutants are hard to determine, as are long-term risks of exposure. Moreover, we have much less adequate information about a host of health effects other than cancer, such as the impacts of environmental exposure on nervous systems, reproduction, and immune systems (Misch 1994).

Ecological Risk Assessment   Much of the debate over risk assessment has focused on health risks. Increasingly, the concepts are being extended to ecological risks (U.S. Environmental Protection Agency 1992a). Because knowledge of ecosystem functioning is even less well developed than knowledge of how pollutants affect public health, we need to ensure that environmental science research can address such questions. Whether the issue is the effects of toxic chemicals on the nation's surface water quality, the impacts of acid precipitation on aquatic ecosystems, or the probability of climate change due to the buildup of greenhouse gases, regulatory decisions are critically dependent on improving environmental science and bringing it to bear more effectively on policy decisions (Carnegie Commission 1993b).

Setting Quality Standards

Once risk assessments permit at least tentative answers to the question of how pollutants affect health and the environment, the EPA and other agencies have to determine the level of contamination that is tolerable based on the defined criteria. These are called *environmental quality standards.* Setting their level involves not only risk assessment but risk evaluation. The latter is a political judgment about how much risk is acceptable to society, or to the particular community or groups at risk, including sensitive populations (such as children, pregnant women, or older persons) and disadvantaged or minority groups. For example, what is the maximum level of ground-level ozone that is acceptable in light of its adverse health effects? How much lead should be allowed in drinking water? What level of pesticide residue on food is tolerable? For years, critics have urged the EPA and other agencies to clearly distinguish between the scientific basis of such choices and the policy judgments and to educate the public and policy makers on the issues. Table 4.1 lists one of the most well developed set of quality standards, the National Ambient Air Quality Standards (NAAQSs) for the six major pollutants regulated by the Clean Air Act.

These crucial decisions about acceptable risk levels are never easy to make even if some analytic tools are available to estimate the public's general risk preferences or tolerances or if the public is able to participate directly in the process. Whatever the basis for decisions, the consequences are important. If environmental and health risks are exaggerated and if unnecessary or excessive regulations are imposed, the nation (or state or community) pays a price in added costs of compliance and possibly wasteful diversion of economic resources. If, on the other hand, risks are ignored or underestimated, we may fail to protect human health and environmental quality sufficiently, and severe or irreversible damage may occur.

As discussed earlier, risk-based priority setting has become a kind of mantra in commentary on reform of environmental policy. The EPA's 1990 study *Reducing Risk* set the tone for this ongoing debate, which is now heard at the state level as well. The EPA is trying to help the states identify and respond to their most pressing

Table 4.1   **National Ambient Air Quality Primary (Health Related) Standards in Effect in 1993**

Pollutant	Averaging time	Maximum concentration[a]
Lead (Pb)	Maximum quarterly average	1.5 µg/m³
Sulfur dioxide (SO₂)	Annual arithmetic mean 24-hour[b]	80 µg/m³ (0.03 ppm) 365 µg/m³ (0.14 ppm)
Carbon monoxide (CO)[b]	8-hour[b] 1-hour[b]	10 mg/m³ (9 ppm) 40 mg/m³ (35 ppm)
Nitrogen dioxide (NO₂)	Annual arithmetic mean	100 µg/m³ (0.053 ppm)
Ozone (O₃)	Maximum daily 1-hour average[c]	235 µg/m³ (0.12 ppm)
Particulate matter (PM-10)	Annual arithmetic mean[d] 24-hour[d]	50 µg/m³ 150 µg/m³

Source: U.S. Environmental Protection Agency, *National Air Quality and Emissions Trends Report, 1992* (Research Triangle Park, N.C.: EPA Office of Air Quality Planning and Standards, October 1993).

[a] Parenthetical value is an approximate equivalent.

[b] Not to be exceeded more than once per year.

[c] Standard is attained when the maximum hourly average concentrations above 0.12 ppm occur no more than one day a year.

[d] Particulate standards use PM-10 (particles less than 10 microns in diameter) as the indicator pollutant. The annual standard is attained when the expected annual average concentration is less than or equal to 50 µg/m³; the 24-hour standard is attained when a concentration of 150 µg/m³ occurs no more than once a year.

environmental risks, recognizing that they lack the financial resources to do everything required by the welter of federal environmental statutes (Stone 1994c). If the comparative risk strategy is to work, methodologies for both human health and ecological risk assessments will need improvement, as will our capacity to evaluate scientific findings and to ensure they are integrated with policy judgments. Risk assessments alone cannot determine policy. And they should not replace careful judgment by scientists, regulators, and the public about proper levels of safety, tradeoffs between risks and benefits, or priorities for regulatory efforts. They can, however, bring much needed information to participants in those decisions.

Emission Standards

Emissions standards follow from the environmental quality standards. They regulate what individual sources (factories, refineries, automobiles, wastewater treatment plants) are allowed to emit into the air, water, or land without exceeding the overall capacity of the environment reflected in the quality standards. In most cases, decisions on air and water permits are made by state agencies using federal standards and guidelines. Such agency decisions reflect judgment about environmental science, public health, available emission control technologies, and, where permitted, economic impacts of regulatory decisions.

BALANCING STATUTORY GOALS AND COSTS

The costs of environmental regulation are high enough that most of the decisions described here have become intensely controversial, whether the issue is automobile emission standards or water quality criteria. The battles often reach the evening news and the front pages of the newspapers. In 1994, for example, the EPA was locked in a heated dispute with the governor of California, Pete Wilson, and the state's farmers over its plan to restore water quality and ecosystem health in the Sacramento–San Juan River estuary and San Francisco Bay. Much of the conflict turned on the validity of the water quality criteria the EPA set for the river and estuaries under the Clean Water Act. As is often the case, some participants asserted that the state had to choose between overzealous environmental protection and the health of the state's agricultural economy. Environmentalists argued that with proper environmental management it was possible to have both high water quality for ecological purposes and sustainable and productive agriculture. The challenge lay in defining the terms of compromise and winning support for them in an atmosphere that did not encourage an open and honest dialogue on the issues (Reinhold 1993b).

One would think from the ferocity of the battles over water use that farmers get little of California's precious supply of fresh water. Yet historically the state's agricultural interests have consumed over 80 percent of the water, much of which has been supplied at heavily subsidized rates to support irrigated pasture and alfalfa for growing cattle. The 1992 Central Valley Project Improvement Act, signed by President George Bush over Governor Wilson's strong opposition, will divert one-fifth to one-sixth of project water to rivers, estuaries, and habitats in the Sacramento–San Juan River Delta to restore fish and wildlife regardless of the EPA water quality rule.

The Costs of Compliance

There isn't much doubt that environmental protection is costly. According to the EPA, the nation spent about $140 billion in 1994 to comply with all regulations written under the seven statutes reviewed in this chapter. Private industry bears about 57 percent of that amount, local governments 24 percent, the federal government 15 percent, and state governments 4 percent (Portney and Probst 1994). The cost is especially striking when viewed historically. The EPA estimates that U.S. spending on pollution control and abatement increased almost fourfold from its 1972 level of $30 billion (in 1990 dollars), or .9 percent of the Gross Domestic Product (GDP), through 1990, when it reached $115 billion, or 1.9 percent of the GDP.

The trend is projected to continue, though at a slower rate of growth. Total spending is expected to reach $171 billion (in 1990 dollars), or 2.6 percent of the GDP, by the year 2000 (U.S. Environmental Protection Agency 1990b). The EPA has several scenarios for future spending levels, depending on whether implementation is at the present level or at a higher "full" level. These numbers reflect the present level of implementation and incorporate the EPA's calculations of annualized costs at 7 percent interest. It should be said that not all of these costs are mandated by federal law. The EPA estimates, for example, that mandated costs for 1990 were $93 billion; for the year 2000, they are expected to be $158 billion. For obvious reasons, long-term projections of environmental spending are subject to large margins of errors.

As these amounts illustrate, compliance costs are a far larger portion of the total than the direct costs of running government programs. Yet their measurement is not as simple and straightforward as industry or state and local governments make out. The cost of complying with environmental regulations depends on technological developments that cannot accurately be foreseen. Annual costs may well decline over time as manufacturing and other industrial processes change. For obvious political reasons, industry tends to use estimates at the high end of the range to argue against additional regulations, whereas environmental groups and government agencies invariably use lower estimates. Estimates of public health and environmental benefits are harder to come by. They also receive less attention than costs in debates over the economics of environmental protection.

Inconsistent Statutory Standards

The statutory language that guides decisions about quality and emission standards could help officials make the tough policy choices. Yet the laws are rarely clear about precisely how risk levels are to be set or how the benefits of regulation are to be compared with costs. As we saw above, statutes like the Clean Air Act and Clean Water Act call for technology-based standards, such as using the "best available technology," or taking all possible steps that technology allows. Others, such as the Delaney clause in the Food, Drug, and Cosmetic Act, call for zero risk. Economists have long complained that such standards create significant inefficiencies in environmental policy by mandating more expensive pollution control than necessary to achieve the desired level of environmental quality (Lave 1981; Freeman 1994; Portney 1990a). They call instead for performance standards, marketable pollution permits, monetary incentives, and information disclosure, among other policy tools, some of which the EPA and the states have experimented with over the years. Environmentalists remain skeptical of those alternatives. They also prefer "no-risk" standards to ambiguous "significant risk" or "unreasonable risk" language to force agencies to achieve the highest level of environmental quality possible.

Statutory specifications in these policies vary because the laws are written by different congressional committees responding to shifting coalitions of interests and changing knowledge of environmental risk. Under some sections of the Clean Air Act, the EPA cannot take costs into account in setting maximum permissible air pollution concentrations. Under the Clean Water Act, in establishing effluent standards, the EPA may consider costs but not benefits. Only two of the seven major environmental statutes—FIFRA and TSCA—require an explicit balancing of the benefits and costs of regulation in setting environmental standards. Critics argue that these specifications help explain why these two acts have been less successful than the other pollution control laws. Yet a study published in 1994 by Resources for the Future indicated that despite the variability in statutory language, there seems to be little practical difference in the way the EPA actually takes benefits and costs into account in drafting environmental regulations (Van Houtven and Cropper 1994). Regardless of how the agency makes these kinds of decisions, the variability in statutory language and requirements has created a degree of complexity in environmental regulation that necessitates elaborate institutional mechanisms and computer databases simply to figure out what the law requires.[10]

FORMULATING AND ENFORCING REGULATIONS

Implementing statutes requires more than setting environmental standards. Agency officials must interpret vague statutory language and develop the means to achieve policy goals. Typically, that requires drafting guidelines and regulations that are legally binding. The 1990 Clean Air Act, for example, is over 400 pages long and requires the EPA to write hundreds of new regulations, 55 of them within two years of enactment. It is scarcely surprising that the EPA falls behind schedule in meeting such expectations, and that it is often compelled to act when environmentalists and others file suit. Such suits have an important drawback. Judicial decisions can force the agency to concentrate on the disputed issues to the detriment of other statutory provisions (Melnick 1983; Wenner 1994).

Administrative Rule Making

The federal Administrative Procedure Act, specifications in individual environmental statutes, and judicial rulings all have sought to make the exercise of administrative discretion open, nonarbitrary, and accountable. They cannot, however, guarantee well-designed or effective regulations. They also provide only minimal assurance of agency responsiveness to public preferences since the administrative process cannot easily accommodate full and regular participation by environmental and citizen groups and certainly not by the general public. Nor can it compensate reliably for the differential access and resources of competing interests.

The process is straightforward in its basic outline, even if its execution is not. When agencies such as the EPA determine a rule is needed, they may publish an advance notice of proposed rule making in the *Federal Register* to signal their intent to consider a rule (although this is unusual). They assemble the requisite scientific, economic, and other data, and then begin formulation of a draft rule or regulation. When a draft is ready, another published notice in the *Federal Register* invites public comment. The agency submits the draft rule to the White House Office of Management and Budget (OMB) for review and clearance, which at times has been done prior to public notice. After consideration of public comments, data, studies, and other material submitted to the agency by interested parties, it publishes the final rule in the *Federal Register,* accompanied by agency responses to the major issues raised during the public participation stage (Bryner 1987). There is a roughly analogous process for state agency rule making.

The EPA and other agencies are required to develop elaborate evidentiary records in a rule-making docket that includes agency planning documents and studies, legal memoranda, advisory committee reports, public comments, summaries of meetings and hearings, scientific and economic analyses, the proposed and final rules, and other pertinent material. The entire rule-making process can easily take three or more years to complete—not counting the time needed for judicial or congressional review (Bryner 1987; Kerwin and Furlong 1992). It comes as no surprise that many participants view the process as cumbersome. EPA insiders, for example, describe the activities of agency work groups that compile scientific and economic analyses and develop the draft regulations as often unwieldy and exceptionally time consuming (Rosenbaum 1994).

These complicated processes are difficult to avoid given the nation's dedication to due process and protection of individual rights. The high stakes involved in environmental policy compound the problem as numerous parties seek to participate in and influence the administrative process. The larger and better financed trade associations, industries, and other organizations employ an army of Washington law firms and technical consultants to help make their case. Although environmental and citizen groups are rarely as well endowed as business interests (Furlong 1992), they often have significant opportunities to shape the outcome, especially at the state and local level.

Participants in rule making bring with them diverse and conflicting perspectives on environmental problems (scientific, legal, administrative, economic, political) that are inherently difficult to reconcile. Unfortunately, the adversarial U.S. political system encourages seemingly endless disputation among them, particularly where procedural delays benefit one or more parties. The disappointed losers have additional options. They can sue the agency and make their case again in the federal courts. Throughout the 1970s and 1980s, about 80 percent of final EPA regulations were contested in court. That figure rises to close to 100 percent for OSHA, which is responsible for some environmental pollutants in the workplace. In recent years, the EPA has been sued about 100 times a year by environmental or business groups or the states. The result can be protracted legal proceedings, long delays in implementing the laws, and excessive costs.

Those effects have prompted EPA administrators to experiment over the past decade with consensus-based rule making, particularly regulatory negotiation (reg neg), in which a committee of affected interests seeks consensus on a proposed regulation. Almost half of the 35 federal negotiated rulemakings completed between 1982 and 1993 or currently in progress have been at the EPA (National Performance Review 1993b). This is a small but nevertheless significant shift from conventional regulatory processes. Although promising, and encouraged by the government since adoption of the Negotiated Rulemaking Act of 1990, regulatory negotiation is not suitable for all situations. It is most appropriate when rule making involves only a few parties who are prepared to negotiate in good faith and who have no incentive to obstruct the proceedings, when the issues do not involve compromise of fundamental principles, and when no great inequities in resources or political power exist among participants (Amy 1987).

Similar efforts at consensus decision making have been emphasized outside of the formal EPA rule-making process, and have been widely endorsed, particularly in disputes over local environmental hazards and land-use issues (Krimsky and Plough 1988; Susskind 1987; Bingham 1986). Al Gore's National Performance Review offered detailed recommendations for promoting greater use of regulatory negotiation and other forms of alternative dispute resolution in federal agencies.

Compliance and Enforcement

Environmental policies mean little without monitoring of compliance by regulated parties and enforcement actions to ensure a high level of compliance. For much of the past two decades, there has been only a halfhearted commitment to both objectives. Monitoring of regulated behavior as well as enforcement have been inadequate, primarily due to the difficulty of the tasks and lack of sufficient resources (Russell 1990). The EPA and the states rely heavily on self-monitoring by industry and other regulated parties, who are to report the results. Studies by Resources for the Future have found infrequent visits and inspections by regulatory officials, the impact of which is limited by the common practice of announcing such inspections in advance (Russell 1990).

The states are responsible for far more water pollution permits and air quality inspections and monitoring actions than is the federal EPA, and there is considerable variation in enforcement activity across the 50 states. States with strong and consistent enforcement actions appear to achieve more success in reducing pollution levels (Ringquist 1993). This finding is consistent with a comparative study of enforcement of effluent regulations for the pulp and paper industries in the United States and Canada that suggests cooperative approaches to regulation may be less effective than traditional "coercive" approaches (Harrison 1995). The last few years have brought much criticism of harsh and inflexible regulations and advocacy of cooperation, negotiation, and use of carrots rather than sticks. Yet we have surprisingly little empirical evidence that speaks definitively to the question of which approach produces the best results.

The vast majority of the time, the EPA and the states rely on negotiated compliance encouraged through informal means such as meetings, telephone conversations, and letters. Only when such efforts fail do more formal means of enforcement come into play. Agencies may then turn to an increasingly severe series of formal actions. These start with Notices of Violation (NOVs), followed by Administrative Orders (AOs), and then formal listing of companies as ineligible for federal contracts, grants, and loans, as allowed by the statute. As a last resort, when all of the above fail to yield compliance, the EPA can move to civil and criminal prosecution with the aid of the Justice Department's Environment and Natural Resources division (Hunter and Waterman 1992). Even here, however, most cases are settled out of court, clearly signaling that the EPA is reluctant to use the courts in routine enforcement actions.

The use of such enforcement tools varies to some extent from one presidential administration to another and even from one EPA administrator to the next. Enforcement actions declined precipitously early in the Reagan administration when Anne Burford ran the EPA, and then increased under William Ruckelshaus and Lee Thomas (Wood 1988; Wood and Waterman 1991). In 1988 there were 372 civil referrals to the Justice Department and 3085 administrative actions for violation of the major environmental protection acts (Russell 1990). Administrative actions and civil and criminal penalties for environmental violations increased steadily in the early 1990s (Council on Environmental Quality 1993).

Because of their visibility and symbolism, criminal prosecutions are sometimes used to signal the government's intent to enforce the law and thus spur voluntary compliance. The CEQ reported that in 1992, the number of criminal indictments handed down reached an all-time high (174), with a record in fines, restitution, and forfeitures as well (much of it, however, from the Exxon *Valdez* case). The Rockwell Corporation agreed to pay the largest RCRA fine in history, $18.5 million, and Chevron pleaded guilty to 65 Clean Water Act violations; it paid $8 million, $6.5 million of which was criminal, the third largest criminal penalty ever assessed under any environmental statute. The laws also hit small businesses, sometimes to set an example. In May 1994 the owner of an automobile repair shop in Saint Louis became the first individual prosecuted under the 1990 Clean Air Act's provisions regulating auto air conditioners. He pleaded guilty to draining auto air conditioners of CFCs without the proper equipment for capturing and recycling the chemical to prevent its release (Wald 1994a). The act provides for up to five years in prison and a fine of $25,000 a day, though such heavy sentences are rarely imposed.

White House Oversight

Concern over the costs and burdens of environmental and other regulations have led all presidents since Gerald Ford to institute some form of centralized White House oversight of agency rule making as part of their broader effort to control the federal bureaucracy. EPA rules have been a major target of the reviews (National Academy of Public Administration 1987; Shanley 1992). For ideological and political reasons, these efforts reached their full flowering under Ronald Reagan and George Bush.

The Reagan Executive Orders   In January 1981 President Reagan formed the Task Force on Regulatory Relief, headed by then Vice President George Bush, to review major proposals by regulatory agencies and to reduce the burden of regulation on business and society. It worked in tandem with a newly created Office of Information and Regulatory Affairs (OIRA) within OMB that was to serve as the key body for regulatory oversight. In addition, Reagan signed two executive orders that guided the process. Executive Order (EO) 12291, issued in February 1981, required the conduct of a formal cost-benefit analysis, or Regulatory Impact Analysis (RIA), prior to formal proposal of major regulations, defined as having an annual impact on the economy of at least $100 million. The order stated further that "regulatory action shall not be undertaken unless the potential benefits to society for the regulation outweigh the potential costs to society." The EPA and other agencies were to select the "alternative involving the least net cost to society." The order authorized OMB to review the documents and enforce the policy. A parallel EO 12498 (issued in January 1985) required the EPA and other federal agencies to develop an annual regulatory agenda for submission to OMB and to indicate how their programs were consistent with the president's own agenda.

Critics faulted the original OMB review process on many grounds, including closed decision making, poor documentation, bias in discussing issues with regulated interests, lack of technical expertise, and regulatory delays (Eads and Fix 1984). To mollify congressional critics (who had threatened to eliminate funding for OIRA), the Reagan administration established public records of decisions and made the review process less overtly political. Yet the process remained hard to defend as a systematic and fair approach to coordinating and improving federal regulation. Critics continued to question OIRA's capacity for judging agency proposals, its ability to review as many as 2400 regulations a year (about 14 percent of which have come from the EPA), and the use of cost-benefit analysis in the manner required by EO 12291, which was biased against approval of new regulations (Cooper and West 1988).

The Council on Competitiveness   President Bush kept both executive orders. When press reports portrayed his administration as engaged in a new round of environmental, health, and safety regulation, he formally established in June 1990 yet another review office to keep regulation in check: a presidential Council on Competitiveness, headed by Vice President Dan Quayle. The Quayle Council was to exercise the same authority over regulation as the Reagan administration's task force, and it soon became embroiled in controversy.

The council quickly became known in the business community for providing a secret "backdoor" in government for industries displeased with agency regulators. Like the early OMB review process, it purposely left no paper trail or "fingerprints" to indicate why it intervened, with whom the staff consulted, and what decisions it reached. The council had a small staff of from five to ten people who worked out of the vice president's office. Though it cooperated with OIRA, it operated independently and sought to enhance the vice president's political reputation by going after what it viewed as particularly outrageous examples of regulatory excess (Berry and Portney 1995). Environmental regulations were a favorite target. For example, the council staff met regularly with EPA staff on the Clean Air Act implementation; in 1991 alone some 47 meetings took place (Duffy 1994). The council urged the agency to make more than 100 changes sought by the business community to EPA air pollution regulations, including some that Congress explicitly rejected in approving the act. It justified its actions as preventing the imposition of "unnecessary costs on the economy" and staying within the $26 billion cost estimate agreed to by President Bush and Congress (Schneider 1992b). The council's activities prompted yearlong congressional hearings and criticism from the CAA's authors, among them Representatives John Dingell (D-Michigan) and Henry Waxman (D-California), that the council was trying to bar effective implementation of the act with actions that were "flagrantly illegal" or, at a minimum, "in clear conflict with the statute" (Bryner 1993, 173–78; U.S. General Accounting Office 1993b; Duffy 1994).

The Effects of Regulatory Constraints   The impact of the antiregulatory fervor of the Reagan and Bush years cannot be assessed rigorously for lack of adequate data. This extends to Bush's order of a 90-day "regulatory moratorium" in January 1992, which was continued in April for an additional four months as the November election approached. A large number of proposed environmental (and other) regulations were reviewed, a few were rejected, and many more (including major rules) were modified significantly along the way (National Academy of Public Administration 1987; Cooper and West 1988). Between 1982 and 1992, the total number of federal regulations did decline appreciably, though mostly in the early years, after which regulatory activity stabilized. There was little change in the number of regulations issued following Bush's moratorium (Furlong 1994). Consistent with Washington practice, the Reagan and Bush administrations asserted that the nation had been spared the expenditure of vast sums of money on needless regulations. The council, for example, claimed that 1992 savings alone under the moratorium may have totaled some $35 billion (Hershey 1992). Rarely was any mention made of the benefits those rejected or modified health, safety, and environmental regulations would have provided to the American public.

On the whole, political rhetoric and symbolism were probably more important than substantive impacts on environmental policy. Whatever regulatory relief was provided to the business community was mostly short term and unaccompanied by genuine and long-lasting reform of public policy and agency procedures. Yet 12 years of executive review did help to institutionalize centralized White House oversight and legitimize reliance on some kind of cost-benefit analysis of environmental regulations. The impact can be seen in the Clinton administration.

The day after taking office, Clinton announced the termination of the Council on Competitiveness. Yet he too recognized the imperative of having some form of White House regulatory review process. On September 30, 1993 Clinton issued Executive Order 12866 on regulatory planning and review. It formally revoked the council's authorization by replacing the two Reagan executive orders, 12991 and 12498. The Clinton EO requires that all written communication between White House staff and outside interests concerning proposed regulations be placed in a public file forwarded to the agency; oral communications are to be noted by date and name of individuals involved. The EO requires timely action by OIRA in reviewing proposed regulations and provision to the issuing agency of a written explanation of all requested changes in a regulatory action. The political culture of the Clinton White House and the tone of the EO suggest that under this new arrangement, there will be a shift of responsibility back toward the agencies, with OIRA reviewing only the most important and far-reaching regulations.

CONCLUSION

With all the criticism directed at it, one might be tempted to conclude that hardly anything was right with the present environmental regulatory regime and only wholesale policy and management change would improve the situation. Such appraisals ignore important, if uneven, improvements in environmental quality over the past two decades that are directly and indirectly tied to environmental policies. They also are oblivious to major shifts now underway in industry and government such as pollution prevention, "green" technology development, and environmental research that arguably are related to two decades of federal environmental policy efforts and strong public support for them.

The problems with environmental policies and with the EPA are real enough, though not without solutions. Some of the blame for the present state of affairs surely lies with EPA and other federal officials who should have done a better job of managing their programs. Congress must share the responsibility for having burdened the EPA with far more tasks than it can possibly handle with its budgetary resources. Depriving the agency of the discretion and tools it needs to set priorities and spend money effectively and efficiently only worsens the problem. So too do the mixed signals the EPA continuously receives from Capitol Hill, the White House, the courts, other federal agencies, state and local governments, environmentalists, and industry.

Thoughtful assessments of what might be done to improve environmental policy are not hard to come by. Solutions, however, depend on policy makers developing a broader and clearer concept of environmental policy goals and the political will to pursue them with effective tools. The American public must be part of that dialogue, admittedly a difficult undertaking in a time of rampant cynicism toward government. The public needs to develop a better understanding of the nature of environmental problems and risks, how they compare, the costs of dealing with them, the tradeoffs involved, and the policy choices we face as a nation.

ENDNOTES

1. Readers who seek a fuller treatment of the statutes and some of the issues raised here should explore the sources cited in the chapter and the edited collection by Paul Portney (1990a). Portney's book offers detailed coverage of U.S. policies on air and water pollution, hazardous waste, and toxic chemicals, with an emphasis on economic impacts. For current developments in environmental policy, two highly useful sources are the Environmental Law Institute's *Environmental Law Reporter* and the Bureau of National Affairs' *Environment Reporter.* The leading environmental law texts offer elaborate reviews of statutory and case law, and some volumes, such as West Publishing's Selected Environmental Law Statutes, reprint environmental statutes from the United States Code.

2. The far-reaching and innovative Southern California clean air plan resulted from precisely such a rejection. The formally submitted California SIP openly admitted that it would not bring the state into compliance with national standards. Local environmentalists sued the EPA, arguing that under the provisions of the Clean Air Act, it could not accept such a plan. A federal appeals court agreed. For the details, see Kraft (1993).

3. For more details on the Clean Air Act, see Bryner (1993) and Portney (1990b). Bryner's account is particularly useful for a comprehensive and informative review of air pollution problems, the major controversies in congressional consideration of the 1990 act, and the political factors shaping its adoption. Portney offers a clear and readable assessment of both statutory provisions and economic aspects of clean air regulation.

4. The history of federal water pollution control dates back to the Refuse Act of 1899, which was intended to prevent constraints on navigation. The act prohibited the discharge or deposit of "any refuse matter" into "any navigable water" of the United States. The volume of waste products such as fiber and sawdust from paper mills and saw mills was so great in some rivers and channels that it threatened to block navigation. Administered by the Army Corps of Engineers, the Refuse Act had little or no impact on most industrial and municipal water pollution (Freeman 1990).

5. For a review of these controversies in 1993, see Schneider (1993c). The issues haven't changed much over the past several years except for the building of additional scientific evidence (reviewed in the 1993 National Academy of Sciences report) in support of restricting pesticide use.

6. Critics said such an accident couldn't happen in the United States. Yet an April 1989 draft report to the EPA found 17 Bhopal-like disasters in the nation over the past 25

years where there was a release of deadly chemicals in volume and at levels of toxicity exceeding those in the Bhopal accident. The report tallied over 11,000 accidents between 1982 and 1988 involving toxic chemicals, which caused 11,341 injuries and 309 deaths. Twenty-nine of these were considered to be "major releases" of toxic chemicals (Shabecoff 1989).

7. Kriz's (1994d) detailed account of these activities is consistent with other reports (e.g., Cohen 1992) suggesting that renewal of the major environmental policies in recent years has depended on extended, behind-the-scenes negotiations between the major adversaries. It is one of the few ways to break legislative gridlock on environmental issues. Private organizations such as the Keystone Center in Colorado have assisted. In early 1993, Keystone, an energy and environmental mediation group, helped pave the way for the Superfund compromise through negotiating sessions with the leading players.

8. Extensive data on the 50 states' environmental conditions and policies can be found in Hall and Kerr (1991).

9. The EPA concluded that dioxin leads to "worrisome" health problems in humans even at extremely low exposure levels, and it was prepared to take steps to reduce exposure to the chemical. Among the health effects are increased likelihood of cancer, damage to reproductive functions, stunted fetal growth, and weakened immune systems. The EPA estimates that dioxin and related compounds are responsible for between 1 in 1000 and 1 in 10,000 of all cancer cases. To add some base of comparison, the risk of smokers developing cancer is between 100 and 1000 times as high. The new risk assessment has taken three years to complete, involved about 100 scientists within and outside of government, and runs to some 2000 pages. The EPA estimated that only about 30 pounds of dioxin are produced annually in the United States (Lee 1994; Schneider 1994c).

10. The EPA has 31 different lists of chemical pollutants that it regulates under environmental laws. The whole business is complicated enough that the agency maintains an Environmental Chemical Listings Information Pointer System (ECLIPS) to run a Register of Lists. Users can enter a chemical's name and discover where in the agency the substance is regulated. In late 1993 the database contained 3475 chemical substances. On the list are 425 carcinogenic compounds, for which information is available on IRIS or the Integrated Risk Information System. Some 922 chemicals are on the CERCLA or Superfund list.

CHAPTER FIVE

Energy and Natural Resources Policy

Much like environmental protection efforts, natural resource policies command significant public support. National consensus reigns on the broad policy goals of natural resource conservation and preservation. Everyone wants to protect treasured national parks and most people would concur in the need to preserve threatened and irreplaceable forests, rivers, or endangered species. Yet conflicts arise over the specific means used to achieve policy goals and on the ways in which competing social values are balanced at any given time and place. How should we weigh the desire to maintain or increase short-term economic growth and employment and the longer-term goal of sustainable use of natural resources? Who should make those decisions? The federal government? State and local governments? Private landowners and free markets?

Such conflicts are familiar in the 1990s, and resemble political battles in other periods in U.S. history. Yet there are important differences. Today, the scale of human intervention is far greater as are the consequences of policy choices. Should we protect most of what little is left of old growth forest ecosystems or sacrifice sizable portions to protect timber industry jobs? Should remaining wetlands be preserved for their vital ecological functions or should we allow development of them? To what extent should we try to reduce reliance on fossil fuels as insurance against the possibility of devastating climate change? In controversies like these, it may be possible to achieve seemingly conflicting objectives through carefully crafted policies of sustainable development. But to do so we will need to devise inventive approaches to decision making that can foster public dialogue, build consensus, and involve major stakeholders without generating environmental gridlock.

These issues have achieved a new prominence as Secretary of the Interior Bruce Babbitt seeks to reverse 12 years of Republican rule that leaned far more toward development interests than the Clinton administration favors. From higher grazing and mining fees for use of federal lands to a comprehensive survey of the nation's biological capital, Babbitt has advocated a natural resources policy agenda that differs dramatically from the goals of the Reagan and Bush administrations. His proposals are more consonant with public sentiment and scientific opinion than are the views of his detractors. Yet the Wise Use and property rights movements add fuel to these policy fires that burn fiercely around virtually every natural resource issue today.

The Wise Use movement is a coalition of groups comprised primarily of timber, mining, ranching, farming, and development interests, some "corporate fronts" and others representing individual property owners concerned about governmental restrictions on use of their property. They operate under the umbrella groups Alliance for America, the National Inholders Association (being renamed the American Land Rights Association), and the Multiple-Use Land Alliance. Together they offer an unabashedly antienvironmental agenda, seeking to weaken the Endangered Species Act and the Clean Water Act's protection of wetlands, promote mineral and energy development in wilderness areas and national parks, and block reform of the Mining Law of 1872, among other actions. The property rights movement, less well organized, has gained momentum since the Supreme Court case of *Lucas v. South Carolina Coastal Council* (1992), in which the justices ruled that in certain circumstances governments may have to compensate landowners when imposing environmental restrictions on development that results in a "taking" of their property.[1]

These continuing disputes over natural resources underscore the difficulty of making collective choices when social values are deeply held and winners and losers can plainly see their fates. This is especially so when the political process is open to pressure from powerful interests eager to protect their share of the nation's economic pie. Scientific studies help to resolve questions of fact, but they cannot substitute for political judgments about where the public interest lies. Even the best studies can build only a partial foundation for environmental policy when policy makers struggle to negotiate a middle course broadly acceptable to all sides.

This chapter provides an overview of these issues in two clusters of environmental problems and policies: energy and natural resources. In each, I review the major policy goals and conflicts as well as the key challenges facing government agencies responsible for policy implementation.

THE VAGARIES OF ENERGY POLICY: GOALS AND MEANS

Energy policy is part environmental protection and part natural resource policy. However, the United States has no explicit and comprehensive energy policy that is comparable to the extensive bureaucratic and regulatory machinery that governs environmental quality and natural resources. Rather, energy use is determined largely by the marketplace, with each major energy source shaped in part through an assortment of government subsidies and regulations adopted over decades for reasons other than energy policy.

Federal and state regulation of coal, natural gas, and oil, for example, has focused on prices and competition within each sector. It has served the interests of energy producers by stabilizing markets and ensuring profits at least as much as it has advanced a larger public interest in reliable energy supplies.

Regulation of nuclear power differs because of its historic connection to national security. From the late 1940s on, under the auspices of the Atomic Energy Acts of 1946 and 1954, federal agencies responsible for the civilian nuclear energy program shielded the technology from the marketplace—and from public scrutiny—to ensure rapid growth of its use. The Atomic Energy Commission (AEC) and its successor agencies, the Nuclear Regulatory Commission (NRC) and the Department of Energy (DOE), vigorously promoted nuclear power as a critical component of the nation's energy mix. Congress contributed as well by subsidizing nuclear energy through restrictions on liability set by the 1957 Price-Anderson Act and provision of lavish research and development funds. In addition, elaborate public relations efforts by federal officials sought to reassure the public on safety and nuclear waste issues. By the late 1970s, following the Three Mile Island nuclear accident, however, public opinion turned against building additional nuclear power plants. It has yet to return to the positive stance of the 1950s and 1960s (Dunlap, Kraft, and Rosa 1993). No utility has ordered a nuclear power plant since 1978.

To the extent that the nation had any discernible energy policy goal before the 1970s, it was to maintain a supply of abundant, cheap, and reliable energy, preferably from domestic sources, to support a growing economy and to ensure a reasonable profit for producers. Policy makers were technological optimists who believed that large, centralized energy sources would meet the nation's needs. For years they were convinced that a growing economy demanded ever increasing energy supplies. Consistent with those beliefs, they championed generous subsidies not only for nuclear energy but for fossil fuels as well.

Increasingly, however, the nation's use of energy has been heavily influenced by environmental policies. It is easy to see why. Exploration for energy sources and their extraction, transportation, refinement, and use degrade the land, air, and water. The 1989 Exxon *Valdez* accident and thousands of lesser oil spills attest to the dangers of moving oil around. Mining coal can ravage the land and poison the waters around mines. Burning oil, coal, and natural gas and their byproducts to generate energy and to power the nation's cars, trucks, and buses pollutes the air, causes acid rain, and leads to the buildup of damaging greenhouse gases. Use of nuclear power produces high-level waste products that must be isolated from the biosphere for thousands of years. Congress has adopted public policies to deal with all of those environmental effects, from clean air and water laws to nuclear waste disposal, surface mining control, and oil-spill prevention measures. By doing so, it has profoundly altered the production and use of energy in important, if sometimes unintended, ways.

Energy policy debate turns on the combination of energy resources that best promotes the nation's long-term interest, and the means that governments use or might use to encourage or discourage the development of particular energy resources. Governments can try to increase energy supply, decrease demand, or alter the mix of fuels used. The choices are not easy, and conflict among diverse parties, particularly regional interests, energy producers and consumers, and environmentalists, frequently leads to stalemate and only modest policy advances. The United States spent an estimated $475 billion for energy in 1992, or nearly 10 percent of its entire economic output. Any governmental actions that substantially alter that spending will have major consequences throughout the economy.

Indirect Policy Impacts

Even without a comprehensive energy policy, federal, state, and local governments currently influence decision making on energy use in myriad ways. They do so through regulation of the byproducts (e.g., air pollution), provision of services (e.g., building highways for motor vehicles), tax subsidies, and energy research and development assistance. The effects of those policies are not neutral and the market is not truly free and competitive. Historically, policies have strongly favored mature and conventional energy sources such as fossil fuels and nuclear power. They also have encouraged expansion of energy use rather than decrease in demand through improved energy efficiency and conservation. A 1992 study by Greenpeace, for example, concluded that taxpayers have spent almost $100 billion (in 1990 dollars) to promote nuclear power since the early 1950s (Wald 1992). A more comprehensive study by the Alliance to Save Energy put a market value of $36 billion on the total federal energy subsidies in 1989 alone, with 58 percent going to fossil fuels and 29 percent to nuclear fission (Koplow 1993).[2]

An example of how such indirect forces work is the use of gasoline (oil) in motor vehicles. The *price* paid at the pump, about $1.15 a gallon for regular in 1994, is not the true *cost* of using the fuel. That amount does not reflect the full expense of road construction and maintenance, national security costs to maintain access to Middle East oil fields, or the environmental damage or externalities related to transporting the oil around the world in supertankers and ultimately burning its gasoline derivative in our cars, vans, and trucks.[3] Yet that cheap market price is all most people see, and it is far too low to stimulate demand for fuel efficient vehicles and mass transit. Because we get used to this artificially low price, and direct tax increases are so visible and resented, policy makers are extremely reluctant to raise gas taxes.

Shifting Energy Priorities

These varied and often invisible government subsidies sanction and stimulate the use of environmentally risky energy sources while erecting barriers to sustainable sources. However, much is changing in the 1990s. Under both the Bush and Clinton administrations, policies and administrative priorities have shifted enough that renewable energy sources, conservation, and improved efficiency are receiving new attention and support in both the public and private sectors. A conversion to sustainable energy sources remains the long-term and widely endorsed goal. The crucial policy questions concern precisely what short-term strategies best promote that goal and also serve short-term needs.

Environmentalists press for an early and rapid shift away from fossil fuels and nuclear power and toward solar, wind, geothermal, biomass, and other forms of renewable energy. They argue that such a transition is possible with well-designed public policies and the political leadership to get them enacted and implemented (Alliance to Save Energy 1991; U.S. General Accounting Office 1993a). Others disagree and find such hopes unrealistic. They see a smaller potential for solar, wind, and other renewables, and more limitations to energy conservation. Thus they suggest a longer-term reliance on fossil fuels and continuation or expansion of nuclear power (Marcus 1992). An examination of recent policy history indicates how these two different visions of a U.S. energy future have shaped policy choices.

THE ENERGY POLICY CYCLE: 1973 TO 1989

At several times over the past two decades, the United States has struggled with defining national energy policy goals, without great success. The early efforts, during the 1970s under Presidents Nixon, Ford, and Carter, were driven largely by concern for the security and stability of the energy supply in the wake of the 1973 oil price shock. In that year, the Organization of Petroleum Exporting Countries (OPEC) imposed an embargo on the sale of their oil, which led to a quadrupling of world oil prices and severe economic impacts. In previous decades, little thought was given to energy conservation, and demand for energy grew dramatically. After 1973, with the reliability of oil supplies called into question, the government made greater efforts to encourage conservation and efficiency of use through a variety of taxation and regulatory actions. These included adoption in 1975 of the Energy Policy and Conservation Act. That act established the Corporate Average Fuel Economy (CAFE) standards for motor vehicles, extended domestic oil price controls, and established the Strategic Petroleum Reserve to stockpile a billion barrels of oil for future emergencies (Goodwin 1981; Marcus 1992).

Carter's National Energy Plan and Conservation Gains

Slowly, energy policy goals shifted to provision of secure and clean energy sources, with reliance on market forces still the preeminent policy approach. President Carter made his National Energy Plan, which he termed the "moral equivalent of war," a top policy priority in 1977 and 1978. A small, ad hoc energy task force working under the direction of Carter's secretary of energy, James Schlesinger, assembled the plan to meet a presidential deadline of April 1977. The group emphasized a strong governmental role rather than reliance on the private sector, and it looked more to conservation than to increasing domestic supplies for solutions to the energy crisis. Preoccupied with producing a comprehensive energy policy analysis consistent with Carter's operating style as president, both the energy planners and the White House neglected to build sufficient political support for their package. They failed to work closely with congressional leaders, energy industry executives, the business community, and environmentalists. They also were unsuccessful in building the broader public support that would be essential to congressional approval.

These flaws in policy legitimation doomed the plan to defeat on the Hill. Congress objected to much in the plan, particularly when its own internal analyses revealed technical flaws that reflected its hasty preparation. Congress also was lobbied heavily by oil and natural gas producing industries, utilities, automobile companies, and labor, consumer, and environmental groups, all of whom found something they disliked in the plan. In the end, Congress enacted some of Carter's proposals and rejected others without substituting an equally comprehensive energy policy.

Congress approved five key components, which collectively were called the National Energy Act of 1978. Among them was the Natural Gas Policy Act, which partially deregulated and altered natural gas pricing to make the fuel more competitive with other sources. Also in the package was the Public Utilities Regulatory Policy Act (PURPA), which helped to create a market for small energy producers using unconventional sources such as solar and geothermal. Other provisions dealt with energy conservation, power plant and industrial fuel use, and energy taxes. Among other actions, Congress approved tax credits for home insulation, energy efficiency standards for home appliances, and taxes on "gas guzzler" cars. Environmental quality was a consideration at the time, but not the main issue. Indeed, Carter's plan aimed to *expand* use of coal because of its domestic abundance despite the environmental consequences of using such a notoriously dirty fuel. Yet Carter also greatly increased support for renewable energy sources.

In 1977 Congress established the Department of Energy, which consolidated previously independent energy agencies into a cabinet department, with the NRC remaining an independent agency charged with overseeing nuclear safety. The DOE's primary mission, however, was not energy but national defense; it was responsible for overseeing nuclear weapons production. Even by 1994, most of DOE's budget supported defense-related activities, including about $5 billion a year for waste management and environmental restoration at contaminated facilities, primarily nuclear weapons production sites.

Despite these policy and institutional limitations, the United States (and other industrialized nations) made great strides in energy conservation in the 1970s and 1980s. Between 1976 and 1986, the United States cut its ratio of energy use to GNP by about 2.8 percent a year (Marcus 1992). Slower economic growth and a decline in older and inefficient heavy industries contributed to the energy savings. Consumer and industry demand for energy-efficient cars, buildings, lighting, motors, and appliances made a difference as well. American industry in the late 1980s used only 70 percent of the energy needed in 1973 to produce the same goods. Appliances in the early 1990s were about 75 percent more efficient than they were in the late 1970s. Passenger automobiles in 1991 averaged about 22 miles per gallon compared to only 14 in 1973, although the average for all motor vehicles in use in 1991 was only about 17 miles per gallon (U.S. Department of Energy 1993). Fuel-efficiency standards for *new* passenger automobiles are 27.5 miles per gallon. The gains are impressive, yet it would be hard to call many of the energy policies of this era a triumph as government programs remained complex, contradictory, and inefficient (Marcus 1992). Except for periods of crisis and following major accidents, energy issues never aroused the public enough to promote a national consensus on overall policy direction or to press Congress to act on the issues.

Reagan's Nonpolicy on Energy

Unfortunately, many of the most innovative policies of the 1970s, including government support for conservation and development and use of alternative energy sources, did not last long. What remained in the early 1980s was cut back sharply during the Reagan administration. Reagan strongly opposed a federal role on energy policy and favored reliance on the "free market." He sought (unsuccessfully) to dismantle the DOE, whose very existence symbolized federal intrusion into the energy marketplace. Some of those positions were broadly endorsed at the time because energy prices were declining. Thus Congress repealed tax breaks for installing energy-saving devices and approved the Reagan budget cuts that effectively ended conservation and renewable energy programs (Axelrod 1984; Rosenbaum 1987). Between 1980 and 1990, support for solar and renewable energy research in the DOE declined by some 93 percent in constant dollars, and the department's energy conservation budget fell by 91 percent between 1981 and 1987.[4] All this was particularly striking because conservation of energy is one of the most effective ways to reduce dependency on imported oil and to lower environmental risks at a modest cost.

The administration's energy budget also concealed fundamental inconsistencies in its ostensible reliance on free-market forces. Support for nuclear programs was increased even while virtually every other program was slashed, with little regard for demonstrable success or failure. Programs to prepare for oil emergencies suffered as Reagan did as little as possible to meet legislative targets for filling the Strategic Petroleum Reserve, leaving the nation vulnerable to oil price shocks. Reagan also ordered his staff to remove solar panels installed at the White House by President Carter, a symbol in conflict with the administration's energy policy agenda. One journalist described the essence of the Reagan strategy as "Duck, Defer, and Deliberate" (Hogan 1984). Even the 55-mph speed limit on interstate highways couldn't survive the new political climate; responding to public pressure, Congress raised it to 65 mph.

ENERGY POLICY FOR THE 1990S AND BEYOND

Energy issues reappeared on the political agenda in the late 1980s as the nation's dependency on oil imports rose once again and concern began to mount about global climate change following the hot and dry summer of 1988. In 1970 the United States imported 23 percent of its oil. By 1991 the figure climbed to 45 percent, thanks to waning domestic energy production, complacency among both the public and policy makers, and increased use of motor vehicles. One reason for the decline in domestic energy production was the stringent requirements of new environmental laws, including the National Environmental Policy Act (NEPA) and the Clean Air Act. NEPA mandated arduous environmental impact statements that made clear the environmental consequences of energy extraction and use, whether the sources were oil, coal, or nuclear power. These effects prompted a debate that continues to this day over whether environmental goals are compatible with the level of energy production the nation has experienced in recent years.

Some two-thirds of the oil burned each day in the nation goes to transportation. Most of it is consumed by private automobiles and trucks. Miles driven per year rose about 70 percent between 1973 and 1993 (Wald 1993). The imports needed to satisfy America's insatiable appetite for oil widened the U.S. trade deficit and further damaged the economy while increasing reliance on a politically unstable region of the world. In 1973 spending on oil imports was $23.9 billion, and the total U.S. trade deficit was $3.9 billion. By 1992, oil imports rose to $44.8 billion, and the trade deficit to $84.5 billion (Wald 1993). Nevertheless, policy makers have been reluctant to press the public to change its energy habits through either higher CAFE standards (regulation) or a steep gasoline tax (market incentives). The effects can be seen in President Bush's struggle to deal with the offshore oil drilling controversy and in congressional reaction to major energy policy proposals by both Bush and Clinton.[5]

George Bush and Offshore Oil Drilling

In response to a 1981 plan by Reagan's secretary of the interior, James Watt, to open a billion offshore acres to oil companies, and in recognition of increasing local opposition to drilling, throughout the 1980s Congress enacted ever broader moratoria on such development. During the 1988 campaign, Bush, a former Texas oil man, agreed to postpone drilling in *certain* fragile areas off the California and Florida coasts. He said he would approve it elsewhere, including within the Arctic National Wildlife Refuge (ANWR) in northern Alaska. By 1990 partisan politics in California and Florida were joined with the drilling issue. Bush sought to aid Republican candidates for governor in both states and protect Republican interests in forthcoming reapportionment battles.

With intense disputes on the issue ongoing within the administration, and strong pressures from the oil and gas industry, Bush set up a task force to study offshore drilling and his options. He agonized over the decision for months, while carefully reviewing a National Academy of Sciences report and other technical studies. A Bush adviser aptly described the dilemma: "This has been a split-the-difference presidency, and on some issues you can't split the difference" (Hoffman 1990). In the end, Bush leaned toward the environmental side. In late June 1990, following the Academy's advice to wait for further study, the president declared a ten-year moratorium for almost 99 percent of California, all of Washington and Oregon, and all of New England north of Rhode Island, including the rich fishing area of Georges Bank. Michael Deland, chair of the White House Council on Environmental Quality, argued that Bush went about as far as any president could under the circumstances. Yet, the oil industry called the decision a "serious mistake" and environmentalists approved only grudgingly. Members of Congress representing other coastal areas insisted that their districts be protected as well, and residents of the protected areas complained that they sought a *permanent,* not a temporary, ban on drilling.

Bush's National Energy Strategy

Some of the same political and economic forces can be seen in Bush's comprehensive energy proposals of 1991, the National Energy Strategy (NES). In 1989 the president had directed the Department of Energy to develop the NES, which it prepared following extensive analysis within the department and nationwide public hearings. The White House substantially revised the NES before sending it on to

Congress in early 1991, displeasing environmentalists who complained that too many energy conservation measures were dropped at the pleading of Bush's economic advisers. The Bush White House conceded that the NES would do little to reduce dependency on imported oil; indeed, it anticipated a further short-term *increase* in dependency. The president said he was committed to the power of the marketplace and that his plan would neither impose new taxes nor increase government regulation. The nation's press portrayed the Bush plan as shortsighted and timid. Even the 1991 Persian Gulf War, fought largely over access to Middle East oil fields (home to two-thirds of the world's known reserves), was not enough to raise public consciousness about energy problems and build crucial support for a strong bill. Polls indicate the public favors tougher conservation measures, yet most people also oppose higher gasoline taxes. The message is not lost on elected officials.

Bush defended the NES as an acceptable balance of energy production and conservation that might help to wean the nation from imported oil by expanding domestic supplies. Environmentalists argued that the plan tilted far too much toward production, risking environmental damage in the process, and that it didn't do much at all for energy conservation or mitigation of climate change. They successfully opposed opening the Arctic wildlife refuge to oil and gas exploration and they pressed hard, but without success, for big increases in the CAFE standards as an alternative to increasing oil supplies. Eventually both the wildlife refuge and CAFE provisions were omitted from the final legislation to gain support from each side.

As is usually the case with energy policy bills, Congress was besieged by lobbyists, particularly from energy interests and automobile manufacturers, who sought to maintain their advantages under current policies. Auto companies, for example, opposed higher CAFE standards on the grounds that doing so would compromise safety by forcing them to build smaller, more dangerous cars. Independent studies, including one by the congressional Office of Technology Assessment (OTA) effectively undercut that argument. The OTA study found that safety would not be compromised providing that auto manufacturers were given sufficient time to develop and use appropriate technologies and that the government would not impose "drastic" increases in fuel economy. In short, better design and safer technologies could offset the risk of lower vehicle weight or smaller size (U.S. Office of Technology Assessment 1991b). Virtually all automobile companies are working on such prototypes. In 1992, for example, General Motors unveiled an "Ultralite" four-passenger car getting 45 miles per gallon in the city and 81 on the highway that also accelerated quickly (0 to 60 miles per hour in 7.8 seconds). GM cut the car's weight to only 1400 pounds without compromising safety (Rocky Mountain Institute 1992).

Political gridlock results when there is little concern and consensus about energy policy among the public and no effective way to restrain the politics of self-interest of energy producers. The effects were evident in the final energy policy bill, which fell far short of the expectations environmentalists hold for national energy policy. Indeed, even the key players on the Hill acknowledged that the bill could be considered a foundation on which to build a more comprehensive energy policy in the future (Idelson 1992a; Kraft 1994a). Nevertheless, the act set out some important goals and established unusual incentives for achieving them. The ultimate impacts will depend on how well the policy is implemented.

The 1992 Energy Policy Act, a massive measure running to some 1300 pages, called for greater energy conservation and efficiency in electric appliances, buildings, lighting, plumbing, commercial and industrial motors, and heating and cooling systems; streamlined licensing requirements for nuclear power plants in the hope of jump starting an ailing industry; and provided tax relief to independent oil and gas drillers to try to stimulate increased production. The act attempted to restructure the electric utility industry to promote greater competition and efficiency by eliminating regulatory barriers that have prevented small producers from gaining access to transmission lines of larger utilities. That provision was expected to carve a place in the market for producers of renewable energy sources such as wind, geothermal, and solar power.

The energy act also required the use of alternative-fuel fleet vehicles by the federal government, which were to be phased in by 1999, when they would constitute 75 percent of federal vehicles purchased. Most state government and private and municipal fleet vehicles are to follow a somewhat less demanding schedule, though Congress wrote in many exemptions to the requirement. California and several Northeastern states will go beyond the federal requirement because their clean air rules mandate use of alternative fuel vehicles by the public as well as state agencies. Other provisions of the act created electric vehicle demonstration programs and public-private partnerships in development of advanced batteries for electric cars and high-efficiency electric trains. Consumers were given generous tax credits (up to 10 percent of purchase price) for buying alternative energy vehicles. Perhaps most importantly, the act authorized billions of dollars for energy research and development, albeit more for fossil fuels and nuclear power than for renewable energy sources (Idelson 1992b).

These are significant achievements, yet problems remain. The market forces that determine energy prices still do not adequately reflect either environmental or national security costs. The act did not raise the CAFE standards for motor vehicles, which was the top priority of environmental groups, nor did it do much to reduce use of oil, which accounts for fully 40 percent of U.S. energy consumption. The alternative-fuel vehicle provisions, while innovative, will have only a minor impact on the nation's oil use. The 1992 energy act should help to slow the growth in U.S. dependency on imported oil, but it is unlikely to either cap or reduce that critical energy dependency.

The Clinton Administration Tries Its Hand

These deficiencies of the 1992 act were apparent as President Clinton assumed office in January 1993. Environmental groups, still unhappy with the 1992 act, urged the president-elect to create a high-level energy task force to review conflicting studies and build political consensus for a "rational" energy policy. Clinton's transition staff promised a series of actions to accelerate energy efficiency in buildings and appliances, government purchase of alternative fuel vehicles and energy-efficient computers, and research support for conservation technologies and renewable energy sources. The president announced in his Earth Day address in April 1993 that he would issue executive orders promoting those actions where possible. As a symbol of his commitment, he also pledged to conduct an energy and environmental audit of the White House to find ways to reduce waste in lighting, heating, and cooling.

In early 1994 Clinton was assembling a larger energy policy game plan to lessen U.S. dependency on imported oil by emphasizing use of natural gas, a domestically abundant and cleaner fuel than oil and coal. Natural gas gives off about 133 pounds of carbon dioxide to generate 100 kilowatt-hours of electricity, oil 169, and coal 210. Hence the basis for preferring natural gas over other fossil fuels in the context of climate change policy. Clinton's energy plan drew broad support from the EPA and the Interior and Energy departments, among other crucial federal agencies (Kriz 1994e).

The president's fiscal 1995 budget continued a shift begun the previous year as he increased funding for energy efficiency and renewable sources at the expense of fossil fuels and nuclear power. Clinton proposed an increase in spending on efficiency and conservation programs of 27 percent, to $743 million, and he boosted research support for natural gas while cutting back sharply on clean coal and nuclear programs. By early 1995 the administration signalled its intention to cut back significantly on overall DOE spending. Still, its fiscal 1996 budget provided modest increases for renewable energy and conservation programs.

Clinton learned early in his presidency that solving the nation's energy problems entailed overcoming serious political obstacles. As part of his deficit-reduction package announced in early 1993, the president had proposed a broad energy tax based on the heat output of fuel or British thermal units (Btu). He expected the tax to raise some $72 billion over five years while simultaneously reducing use of fossil fuels, and thus curbing pollution and reducing oil imports. The proposal bore some resemblance to the carbon tax (based on carbon dioxide emissions) long advocated by environmentalists.

Clinton's Btu tax was an instant political failure. It was greeted by a tidal wave of opposition on Capitol Hill, reflecting complaints from industry groups and others (farmers, energy-producing states, and states reliant on home heating oil) that the plan would cost them too much money, reduce their competitiveness, and put people out of work. The National Association of Manufacturers, the U.S. Chamber of Commerce, the Chemical Manufacturers Association, and the American Petroleum Institute, among others, worked actively to defeat the Btu tax. The antitax forces mounted what one journalist called a lobbying "juggernaut," that "whipped up a froth of outrage over the tax" through sophisticated use of satellite feeds, talk radio, opinion polls, a blizzard of newspaper editorials, and mass mailings to citizens urging them to protest the tax (Wines 1993). The White House did little to defend its energy proposal, and it was forced to retreat step by step as it granted exemptions to one powerful industry group after another. Eventually only a hollow shell of an energy tax remained.

Much to the disappointment of the environmental community, Max Baucus (D-Montana), chair of the Senate Environment and Public Works Committee, announced his opposition to any gas tax that would impose what he viewed as an unfair burden on large states where residents drive long distances. Baucus objected even to a 6.5 cents-a-gallon tax increase. Ultimately, House and Senate conferees agreed to a minimal increase in the federal gas tax of 4.3 cents-per gallon, despite the lowest market price for gasoline in a generation (Wald 1993). Gas taxes in the United States remain far below those of other industrialized nations. In 1992, for example, regular gasoline in the United States sold for about $1.22 a gallon, with federal, state, and local taxes accounting for $0.34 of the total. The same gallon in most European nations sold for about $3.60, with taxes ranging from about $1.60 to $3.50 per gallon. The politics of energy in the 1990s dictated that only painless measures would be acceptable in the U.S. Congress.

The administration's Clean Car Initiative, recently renamed the Partnership for a New Generation of Vehicles, may help to achieve some of the same objectives as the rejected Btu tax. The federal government is to coordinate its R&D spending by seven agencies (which totals about $300 million a year) with Ford, Chrysler, and General Motors, which have pledged a comparable amount. The aim is to produce a car that can go 80 miles on a gallon of gasoline without a loss in performance, passenger capacity, or safety. The Big Three automakers have been somewhat reluctant partners, openly criticizing state policies calling for sale of electric vehicles. The companies are more positive about other fuel sources, including fuel cells, flywheels, and capacitors used in vehicles built with lightweight yet strong materials. Critics have complained that smaller and more innovative companies have been excluded from the research subsidies to date, though the administration promised to include them in the future (Wald 1994c).

State and Local Energy Initiatives

Some of the most promising energy policy initiatives occur outside of Washington, D.C. State and local policy makers can more easily build consensus for innovations than is possible in the contentious arena of national politics. A good example is Sacramento, California, where residents voted in 1989 to close their publicly owned but troubled Rancho Seco nuclear power plant that provided half of the local power. The Sacramento Municipal Utility District (SMUD) is now thriving as a laboratory for energy conservation and use of renewable fuels. It replaced the nuclear plant with natural gas units (and plans others that will use co-generation technology), temporarily buys excess power from neighboring utilities, and expects the rest of its future capacity needs to be met entirely from conservation and renewables. Like many utilities today, SMUD pays homeowners to turn in old, inefficient appliances and to add home insulation. It also has planted tens of thousands of trees to reduce cooling needs, all of which is cheaper than building new generating facilities (Lippmann 1993).

The pattern is not restricted to Sacramento. Tight environmental restrictions throughout California make conservation and efficiency highly attractive to utilities in the state that now find it difficult or impossible to build additional generating plants. Some of the same forces are emerging in states such as Minnesota, Wisconsin, and Colorado. Helping customers buy more efficient appliances and conserve energy (so-called demand-side management or DSM) is the equivalent of constructing new plants. In California these conditions, along with astute political and corporate leadership, a collaborative decision-making process, and energy markets dramatically altered by the federal PURPA of 1978, have created an "energy-environmental revolution." As evidence of its effect, between 1973 and 1988, the state's population increased by 37 percent and its economy by 46 percent, yet energy consumption went up by only 8 percent (Mazmanian 1992). California utilities remain in the forefront of private sector development of alternative energy sources. Southern California Edison already gets 13 percent of its electricity from renewable sources compared to only 1 percent in 1985. By late 1993, these technologies were enjoying a renaissance nationwide, which would be accelerated with additional federal support (Regan 1993b).

NATURAL RESOURCES AND POLICY CHANGE

Equally significant changes are taking place in U.S. natural resources policy in the 1990s. They are evident in nearly every natural resource controversy in Congress, from Interior Secretary Babbitt's plan to reform federal land policy to renewal of the symbolically important 1973 Endangered Species Act. A laissez-faire stance on ecologically damaging activities and generous government subsidies for shortsighted and uneconomic resource extraction are slowly giving way to the goal of sustainable development and the use of ecosystem management approaches. The short-term conflicts get much attention, but the real news lies in the redefined resource agenda and new public values driving environmental policy within the nation and globally.

Environmental Stewardship or Economic Development?

The United States is richly endowed with natural resources. Even though the federal government gave away much of its original land before the early twentieth century conservation movement curtailed the practice, the public domain includes over 740 million acres or one-third of the nation's nearly 2.3 billion acres of land. About half of that is in Alaska and much of the rest in the spacious western states (Wengert 1994). These public lands and waters include awe-inspiring mountain ranges, vast stretches of open desert, pristine forests, spectacular rivers and lakes, and the magnificent national parks: Yellowstone, Yosemite, the Grand Tetons, the Grand Canyon. They also contain valuable timber, minerals, energy resources, and water vital to irrigated crops in the West. Even submerged offshore lands are precious. The federally governed 1 billion acres of land on the Outer Continental Shelf (OCS) contain an estimated 40 to 60 percent of the nation's undiscovered oil and natural gas reserves.

   Congress chose to set aside some areas—the parks, wild rivers, wilderness areas, and national seashores—to protect them from almost all development. Most public land is not fully protected. Rather it is subject to long-standing, but intentionally vague, "multiple use" doctrines that Congress intended to help balance competing national objectives of economic development and environmental preservation. Agency administrators are expected simultaneously to protect and exploit resources. They preserve public lands and waters for recreation and aesthetic enjoyment and protect ecologically vital watersheds and fish and wildlife habitat. Yet they also try to ensure commercial development of the commodities on those lands. Thus agency officials serve as stewards of the public domain and referee disputes among the multitude of interests competing for access to it.

Natural resource policies govern those decisions, most of which fall within the jurisdiction of the Interior and Agriculture departments. Conflict over them escalated dramatically over the past 30 years with the rise of the environmental movement, rapid population growth in the West, and surging interest in recreation, all of which created new demands on the public lands. Added to these developments were structural economic shifts that have imperiled extractive industries like forestry and mining and threatened people whose livelihoods depend on continued access to public lands and waters. As the Wise Use movement of the 1990s vividly demonstrates, crafting solutions that satisfy all parties is always difficult, and sometimes impossible. How well we are able to build such consensus speaks to the nation's capacity to put the concepts of sustainable development and ecosystem management into practice. Success will depend on generating credible and compelling scientific analyses of the environmental effects of resource use and designing policies to mitigate unavoidable adverse economic and social impacts. Few people advocate having crucial value choices made by a centralized and distant bureaucracy. Thus equally critical to the goal of sustainable development is the creation of decision-making processes capable of fostering constructive policy dialogue and consensus building among stakeholders at all levels of government, and especially in affected communities.

The Environmentalist Challenge to Resource Development

Natural resources policy has a much longer history than either environmental protection or energy policy. As discussed in Chapter 3, policies designed to encourage settlement of the West transferred over 1 billion acres of federal domain land to the states and private parties before being ended officially in 1976. Well before the environmental decade of the 1970s, the conservation movement instituted the first of a series of protectionist policies with creation of national parks and monuments and establishment of the National Park Service, Forest Service, Bureau of Reclamation, and other federal resource agencies. For most of the past century, conflicts over protection and development of federal lands were relatively muted and politically contained, in part because natural resource subgovernments, as discussed in Chapter 2, dominated the issues and promoted consensus on policy goals and means (Culhane 1981). Scholars disagree about the extent of agency "capture" by the regulated interests. Yet it would be fair to say that long-settled policies that governed mining, logging, grazing, agriculture, and other uses of public lands and waters rarely emerged as major political issues with the larger public. All that changed as the environmental movement gathered steam in the 1960s. It scored its first policy successes in resource conservation.

Gains in Resource Protection and Political Access   The Wilderness Act of 1964 created the National Wilderness Preservation System to set aside undeveloped areas of federal land where "the earth and its community of life are untrammeled by man." The Land and Water Conservation Fund Act of 1964 provided federal grants to the states for planning for, acquiring, and developing land and water areas for recreation. The fund also provides money to buy property for national parks, forests, and refuges managed by the federal government. In 1968 Congress established the National Wild and Scenic Rivers System to protect free-flowing rivers in their natural state. In a fitting symbol of the dozens of environmental measures to follow, on the last day of 1969, Congress enacted the National Environmental Policy Act, requiring the preparation of environmental impact statements (EISs) for all "major Federal actions significantly affecting the quality of the human environment." Of equal importance was the creation by NEPA and other statutes and judicial rulings of the 1970s of a major role for the public in environmental decision making, altering forever the political dynamics of natural resources policy (Dana and Fairfax 1980).

These new policies, and the political movements that inspired them, brought to the fore the previously latent conflicts in natural resources that are now so visible. During the 1970s, environmentalists gained access to the natural resource subgovernments, from the national to the local level. They became regular participants in agency decision making, and they gained powerful allies on Capitol Hill as natural resource committees and subcommittees began to reflect the public's strong support for environmental protection efforts. They also played an active role in the courts, using NEPA's environmental impact statement process to oppose many environmentally damaging development projects.

Reaction in the West   None of this was good news for traditional resource constituencies, particularly in the western states. They saw their historic access to public lands and waters jeopardized by the new demands for preservation and recreation and by a different set of policy actors unlikely to acquiesce to the old distributive formulas. In the early 1980s, and more recently under the Wise Use banner, the user groups put forth arguments couched in terms of grand political principles: centralization versus decentralization of governmental authority, federal dominance and states rights, property rights and their infringement by government regulations. These are unquestionably important issues, and legitimate differences exist over them. But at heart the political disputes reflect the disaffection of traditional beneficiaries of federal largesse. Their resentment grew throughout the 1970s, and it culminated in the Sagebrush Rebellion (see below) that so influenced Ronald Reagan's environmental agenda and the determined efforts by his first secretary of the interior, James Watt, to reverse two decades of progressive natural resource policies (Culhane 1984; Leshy 1984; Kraft 1984).

Ultimately, Watt and his successors at Interior through George Bush's presidency—William P. Clark, Donald P. Hodel, and Manuel Lujan Jr.—changed few formal resource policies that could not be undone by Babbitt or other administrators.[6] The major policies and programs continue to reflect the ambivalence the nation displays toward the use of natural resources. Strong public support exists for resource conservation and environmental protection, although the issues often fail to command the visibility needed to mobilize the public. The result is that consensus in Congress on these statutes is less than robust, and members tend to be sensitive to the pleas of politically significant constituencies. Mining, logging, and agricultural interests lobby intensely to protect their benefits, and they are well represented on the Hill. This is particularly so in the Senate, where sparsely settled western states enjoy the same representation as the most populous states.

These political forces help to explain congressional opposition to Babbitt's policy proposals, but at least as important is the absence of public concern. Few people not directly affected by the policies are likely to think much about alternative proposals for reform of the Mining Law of 1872 or appropriate fee structures for grazing cattle on federal land. The low salience of the issues weakens the environmentalists' ability to push their reform agenda and contributes to the legislative gridlock that has characterized these policy conflicts in recent years.

Ironically, the West is being transformed politically by an enormous influx of "lifestyle refugees" into Montana, Idaho, Colorado, Oregon, Washington, Nevada, Utah, New Mexico, and Arizona. The new migrants have created a more environmentally oriented electorate, which may have been a contributing factor in Clinton's victory in 1992 in seven of the western states, including five in which George Bush had won in 1988. Despite Bush's efforts on behalf of the timber industry in the Pacific Northwest, in 1992 Clinton won with very large margins (at least 10 percentage points) in all three states in the region: California, Oregon, and Washington. These demographic shifts already have affected Congress as noted above, and they are likely to have a more significant effect over time.

NATURAL RESOURCE POLICIES AND AGENCIES

Much like federal energy policy, natural resource policies are both simple and complex. They are simplest at the level of basic choices made about preservation or development of public resources or equity in the payment of user fees. They are most convoluted in the detailed and arcane rules and procedures governing program implementation. Debates over policy proposals similarly can be fairly straightforward or all but indecipherable to those outside the resource policy communities.

Table 5.1 offers a brief description of the major federal natural resource policies and lists their implementing agencies.[7] To understand the politics of natural resources policy requires at least some attention to the agencies themselves. Congress has granted the agencies enormous discretion to interpret and implement the statutes, which puts them at the center of political battles over protection versus economic exploitation of the public domain at a time of fundamental policy and administrative changes. The conflicts can be illustrated with a selective review of how the agencies are implementing the new resource policies.

Administration of public lands is assigned chiefly to four federal agencies. The Bureau of Land Management, Fish and Wildlife Service, and National Park Service are housed in the Interior Department. The Forest Service, often at odds with the Interior Department, has been in the Department of Agriculture since Gifford Pinchot had the nation's newly created forest reserves transferred there in 1905. Together the four agencies control more than 625 million acres (about a million square miles), or over 90 percent of the total public domain lands. The remaining lands fall under the jurisdiction of the Department of Defense, other Interior and Agriculture department

Table 5.1   Major Federal Natural Resource Policies

Statute	Implementing Agency	Key Provisions and Features
Multiple Use–Sustained Yield Act of 1960, PL 86-517	Interior Department; Forest Service	Defined multiple use and sustained yield concepts for the Forest Service. Required that officials consider diverse values in managing forest lands, including protection of fish and wildlife. Made clear that the forests should be managed to "best meet the needs of the American people," and not necessarily to yield the highest dollar return.
Wilderness Act of 1964, PL 88-577	Agriculture Department; Interior Department	Established the National Wilderness Preservation System comprised of federal lands designated as "wilderness areas," to remain "unimpaired for future use and enjoyment as wilderness." Authorized review of public lands for possible inclusion in the system.
Land and Water Conservation Fund Act of 1964, PL 88-578	Interior Department	Created the Land and Water Conservation Fund, which receives money from the sale of offshore oil and gas leases in the Outer Continental Shelf. Authorized appropriations from the fund for matching grants for state and local planning, acquisition, and development of land for recreation purposes and for acquisition of lands and waters for federal recreation areas.
Wild and Scenic Rivers Act of 1968, PL 90-542	Interior Department	Authorized protection of selected rivers with "outstandingly remarkable features," including scenic, biological, archaeological or cultural values. Rivers may be designated as wild, scenic, or recreational.
National Environmental Policy Act of 1969, PL 91-190	All agencies; coordinated by CEQ	Declared a national policy to "encourage productive and enjoyable harmony between man and his environment"; required environmental impact statements; created Council on Environmental Quality.
Marine Protection, Research, and Sanctuaries Act of 1972 (Ocean Dumping Act), PL 92-532	EPA	Authorized research and monitoring of long range effects of pollution, overfishing, and other acts on ocean ecosystems. Regulated ocean dumping through an EPA permit system that allows disposal of waste materials only in designated areas.
Coastal Zone Management Act of 1972, PL 92-583	Office of Coastal Zone Management	Authorized federal grants to states to develop coastal zone management plans under federal guidelines and to acquire and operate estuarine sanctuaries.
Marine Mammal Protection Act of 1972, PL 92-522	National Marine Fisheries Service	Established a moratorium on the taking of marine mammals and a ban on importation of marine mammals and products made from them, with certain exceptions. Created a federal responsibility for conservation of marine mammals.

Table 5.1   *Continued*

Statute	Implementing Agency	Key Provisions and Features
Endangered Species Act of 1973, PL 93-205	Fish and Wildlife Service; National Marine Fisheries Service	Broadened federal authority to protect all "threatened" as well as "endangered" species; authorized grant program to assist the states; required coordination among all federal agencies.
Federal Land Policy and Management Act of 1976, PL 94-579	Bureau of Land Management	Gave Bureau of Land Management authority to manage public lands for long-term benefits; officially ended policy of disposing of public lands through privatization. Provided for use of national forests and grasslands for livestock grazing under a permit system.
National Forest Management Act of 1976, PL 94-588	U.S. Forest Service	Extended and elaborated processes set out in the 1974 Forest and Rangeland Renewable Resources Planning Act. Gave statutory permanence to national forest lands and set new standards for their planning and management, including full public participation; provided new authority for management, harvesting, and selling of timber, and restricted timber use to protect soil and watersheds; limited clearcutting.
Surface Mining Control and Reclamation Act of 1977, PL 95-87	Office of Surface Mining	Established environmental controls over surface mining of coal; limited mining on farmland, alluvial valleys, and slopes; required restoration of land to original contours.
Alaska National Interest Lands Conservation Act of 1980, PL 96-487	Interior Department; Agriculture Department	Protected 103 million acres of Alaskan land as national wilderness, forest, wildlife refuges, and parks, and other areas of special management and study.

agencies, the Department of Energy, and the Tennessee Valley Authority (Council on Environmental Quality 1993).[8]

As Clarke and McCool (1985) demonstrate, each agency has its distinctive origins, constituencies, characteristics, and decision-making style. The Fish and Wildlife Service and National Park Service lands are governed by specialized missions and for the most part are well protected from development. In contrast, Forest Service and Bureau of Land Management lands are subject to multiple-use doctrines that pose a greater risk of environmental degradation. They also require agency administrators constantly to juggle competing interests and reconcile conflicting interpretations of the law, always under the watchful eye of Congress. The two agencies, however, have markedly different histories and orientations (Dana and Fairfax 1980).

Managing the Nation's Forests

Forests occupy about one-third of the U.S. land area, with the majority owned privately or by the states. The nation has been losing half a million acres of private forestland a year to urban expansion and agriculture, which contributes to ardent interest in those forest lands under federal control (Council on Environmental Quality 1993).

The responsibility for managing the federal lands and the Forest Service itself reflect the era when both were born: Pinchot's progressive conservation movement at the turn of the century. Its awkward blend of resource protection and use continues to this day. The Forest Service itself has a long tradition of professional forest management though environmentalists often have faulted it for excessive devotion to the interests of the timber industry. They have been particularly critical of the service's approval of clearcutting of forests and its insufficient protection of habitat and biological diversity (Culhane 1981).

At the instigation of the Forest Service and to respond to an increasing number of people seeking to enjoy the trails and streams of the national forests, Congress enacted a milestone piece of legislation in 1960, the Multiple Use–Sustained Yield Act (PL 86-517). It defined multiple use as including outdoor recreation, fish and wildlife, and range as well as timber production, and made clear that the forests should be managed to "best meet the needs of the American people," and "not necessarily [through] uses that will give the greatest dollar return." The act set no priorities, and thus gave the service discretion to expand its preservation of forest resources.

The National Forest Management Act   Environmentalist concern rose significantly as the Forest Service began implementing the National Forest Management Act of 1976 (NFMA) early in the Reagan administration. The NFMA amended the 1974 Forest and Rangeland Renewable Resources Planning Act, which had set timber production as a primary goal. But the 1974 act also had established a general planning process that helped to shift the Forest Service (and the Bureau of Land Management) away from what had been an overwhelming emphasis on timber production.

The 1976 act extended this process and added further specifications. It required the Forest Service to prepare long-term, comprehensive plans for the lands under its jurisdiction, and to involve the public in its decision making through meetings and hearings. Thus the process made explicit the tradeoffs between protecting the forest environment and allowing commercial development of it under the legal doctrines of sustained yield and multiple use.

Conflicts between forest preservation and development became especially acute as the service began to increase allowable timber harvest substantially in the mid-1980s. Annual harvests doubled between 1982 and 1989 as the timber industry sought more lumber to meet rising public demand, and the service pursued that goal (Rosenbaum 1991). Environmentalists and industry, or both, regularly contested the forest plans and environmental impact statements that identified such anticipated cutting. Under an 85-year-old rule governing timber sales, environmentalists also have been able to file administrative appeals that delayed or blocked the Forest Service from proceeding with particular sales. The number of such challenges increased from 133 in 1987 to 636 in 1991, prompting the Bush administration to try to limit the rule to curtail the practice as part of administration's election-year assault on environmental regulation and its appeal for votes in the West. The Congressional Research Service reported in 1992 that in one-third of these challenges to timber sales, environmentalists were able to demonstrate that the sale would cause severe ecological harm, leading to decisions by the Forest Service to withdraw or reevaluate the sales (Schneider 1992a). The controversy over cutting of old growth forests in the Pacific Northwest is merely one example of these broader conflicts. Even with such disputes over forest plans, the planning process was widely considered a model for natural resource management (Clarke and McCool 1985, 42).

Changes in the U.S. Forest Service   The Forest Service is a large agency, with over 40,000 employees, a budget of $3.5 billion, and responsibility for over 190 million acres of public lands. In 1991 it sold 6.4 billion board feet of timber (half of the 1989 level), bringing $1 billion in revenues to the federal treasury. Critics have long asserted, however, that the total revenues from timber sales are deceptive because the costs incurred by the government in providing access for timber harvesting (e.g., building and maintaining logging roads) sometimes exceed the revenue. Hence references to "below-cost" timber sales. Even the Forest Service concedes that sales of timber in over half of the nation's forests lose money, and critics complain that the prices charged are often lower than in timber sales on nearby private land.

The Forest Service also supervises grazing, use of water resources, and recreation on its lands, and it does so with a strong sense of mission and high professional standards. In their assessment of federal resource agencies, Clarke and McCool (1985) award the Forest Service one of two "bureaucratic superstar" ratings (the other goes to the Army Corps of Engineers) based on reputation for bureaucratic power, professionalism, and political acumen.

Such accolades notwithstanding, like all resource agencies, the Forest Service in recent years has been undergoing significant changes as it adjusts to new expectations for sustainable resource management and ecosystem protection. At the end of the Reagan years, disaffected foresters formed an Association of Forest Service Employees for Environmental Ethics to press for greater emphasis on environmental values and norms of stewardship through the so-called New Forestry. More than 1000 employees joined within several years of its founding. About the same time, the Forest Service developed a New Perspectives program incorporating some elements of ecosystem management. Studies by the National Research Council (1990), among others, have pressed the service to move further in that direction. Those studies encourage placing greater emphasis on the ecological role of forests and their effects on climate change, educating the public on such issues, and financing Forest Service activities with greater reliance on fees from all users, from loggers and miners to campers and hikers. Such fees would reduce pressures to favor logging (O'Toole 1988).

In 1993 the Clinton administration named wildlife biologist Jack Ward Thomas, leader of a major Forest Service study group on the northern spotted owl, as chief of the service. He promptly informed his top staff that among the major principles he expected to govern their decisions was the use of ecosystem management, building and synthesizing new scientific knowledge to be used in forest management, and establishing public trust in the agency. In 1994 thousands of Forest Service employees tied to its timber production programs took early retirement while the service tried to retain the biologists and recreational specialists essential to new priorities.

Battles Over Wilderness

The history of wilderness designation demonstrates many of the same conflicts of values over whether public lands should be set aside in a protected status or commercial development permitted. For many, these are economic decisions, incorporating a view of natural resources as commodities to be exploited. Others see the decisions primarily as involving moral choices in which they demand the greatest form of protection possible for the few wild places remaining in the nation. Government policy makers are obliged to follow legal dictates in their decision making, but they are pulled regularly in one of these two directions as they exercise the discretion given to them under the law.

As part of its review of land use in the 1970s, the Forest Service embarked on an ambitious attempt to inventory roadless land within the national forest system for possible inclusion in the protected wilderness system. The first Roadless Area Review and Evaluation survey (RARE I) was completed in 1976. It elicited strong disapproval by environmental groups over inadequate environmental impact statements and insufficient public involvement in the process. The Carter administration completed a second survey, RARE II, in 1979. It recommended more wilderness areas, but also sought to meet public demand for timber. More litigation by environmentalists and the states followed, again based on insufficient adherence to NEPA procedures.

The Reagan administration was far more skeptical about the value of wilderness preservation, and its actions reflected those beliefs. Interior Secretary Watt provoked enormous controversy over proposals to open wilderness areas to mineral development (which the Wilderness Act allowed under some restrictions). Watt's actions led Congress in 1982 to order a moratorium on such mineral leasing, which Watt himself extended to areas under consideration for wilderness designation. Yet the administration continued to oppose extension of wilderness areas.

After a 1982 federal court of appeals ruling that some portions of RARE II failed to meet NEPA's environmental impact statement requirements (but not for those lands recommended for wilderness designation), the Reagan administration ordered a broad RARE III on *all* lands on which Congress had not yet acted. Congress had been moving ahead on an ad hoc basis to set aside wilderness tracts and to impose its own priorities for such land use, generally in opposition to the Reagan administration. In early 1983 Reagan directed that RARE III begin. He ordered an acceleration of efforts to eliminate procedural obstacles to development in possible wilderness areas, and Watt withdrew a million acres of potential wilderness from additional study. Those actions further irritated Congress and wilderness advocates. Reagan and Watt persistently misjudged the public and congressional mood on wilderness issues. With its intransigence on wilderness areas, the administration managed to undercut its other natural resources policy initiatives as well (Leshy 1984).

The biggest congressional designation of wilderness areas came before the Reagan era, in 1980, with the creation of 50 million acres of wilderness in Alaska, much of it on Bureau of Land Management land. In 1990, after years of environmentalist pressure and resistance by Alaska, Congress withheld from logging more than 1 million acres of the Tongass National Forest in Southeast Alaska. Nearly 300,000 acres were set aside as wilderness in this last major expanse of temperate rain forest in North America, and limited mining and road building is to be allowed on over 700,000 acres.

Governing the Range

The Bureau of Land Management (BLM) has leaned even more heavily than the Forest Service has toward the resource use end of the spectrum. The BLM governs more federal land than any other agency, but it has suffered from a historic indifference to environmental values compared to the Forest Service. The BLM has a budget that is only about one-third that of the Forest Service (at $1.07 billion in fiscal 1994) and a staff one-quarter the size (about 8700), far less than what is needed to implement its programs. As a result of the Alaska Lands Act in 1980, the bureau was given new responsibility for vast acreage in Alaska, including large areas of wilderness, that now constitutes one-half of BLM land. The bureau is also in charge of federal mineral leases on all public domain and OCS lands.

BLM Deficiencies and Needs   Clarke and McCool classify the BLM as a "shooting star" agency that burned brightly for a short time but now faces a precarious future. It is a relatively new federal agency, formed only in 1946 by executive actions merging Interior's General Land Office and the U.S. Grazing Service, two agencies with a reputation for inept land management. The BLM is widely viewed as weak politically and historically highly permissive toward its chief constituency groups, the mining industry and ranchers. Those characteristics have led scholars to term it a "captured" agency (Foss 1960; McConnell 1966).

As Paul Culhane (1981) observed, however, in more recent years the BLM has recruited professional staff with credentials in scientific land management and a commitment to progressive conservation comparable to those in the Forest Service. The difference between the two agencies lies in the political milieu in which they function. To meet new expectations for sustainable resource management on public lands, the BLM will need to gain more political independence from his traditional constituencies and build a broader base of public support.

The Federal Land Policy and Management Act   Congress tried to stimulate such an organizational shift with passage of the Federal Land Policy and Management Act of 1976 (FLPMA), also known as the BLM organic act. FLPMA formally ended the 200-year old policy of disposing of the public domain, repealed more than 2000 antiquated public land laws, amended the 1934 Taylor Grazing Act, and mandated wilderness reviews for all roadless BLM lands with wilderness characteristics. More importantly, the act legislatively established the bureau as an agency and gave it the authority to inventory and manage the public lands under its jurisdiction. Agency officials had sought the authorizing legislation to clarify its responsibilities and to give it the legal authority to apply tools of modern land management. The new legislation helped the BLM establish greater authority over the public lands under a broad multiple-use mandate that leaned toward environmental values and away from a position of grazing as dominant use.

FLPMA gave the BLM full multiple-use powers that matched those of the Forest Service. It defined multiple use in a way that should encourage environmental sustainability:

> [M]anagement of the public lands and their various resource values so that they are utilized in the combination that will best meet the present and future needs of the American people . . . a combination of balanced and diverse resource use that takes into account the long-term needs of future generations for renewable and nonrenewable resources, including, but not limited to, recreation, range, timber, minerals, watershed, wildlife and fish, and natural scenic, scientific and historical values. . . .

Like most legislation, FLPMA represented a compromise needed to secure the approval of key interests (Dana and Fairfax 1980).

Much like its counterpart governing the nation's forests, FLPMA established a land-use planning process for BLM lands that required extensive public participation and coordination with state and local governments and Indian tribes. Here too, Congress specifically chose to delegate authority for land-use decisions to the agencies. The act increased the authority of the BLM to regulate grazing, allowing the secretary of interior to specify the terms and conditions of leases and permits, including reduction in the number of livestock on the lands. But it also maintained the grazing permit system, the local boards that supervise it, and the existing fee structure. All have been sharply criticized by study commissions and environmentalists for contributing to the extensive degradation of public rangelands (Henning and Mangun 1989).

BLM Lands and the Sagebrush Rebellion

Since the passage of FLPMA, the BLM has been moving slowly toward the new goal of sustainable resource management. Indeed, the Sagebrush Rebellion of the late 1970s largely grew out of the frustration of western ranchers angry over BLM's implementation of that law. Their reaction was stimulated by decisions concerning grazing fees and protection of wildlife, native plants, and predators (Clarke and McCool 1985; Durant 1992). The ranchers tried to force the transfer of federal lands to state or private ownership where they believed they could exert more control. In 1979 the Nevada legislature approved a bill demanding that the federal government turn over all BLM land in Nevada to the state, an action usually credited as the first shot fired in the rebellion. Eighty-two percent of Nevada is federally owned, with smaller, but still large, percentages common in the western states (running from 28 to 68 percent).

Four state legislatures followed in Nevada's path, but most western governors were publicly or privately opposed to the action. Apparently, most people in the western states did not support the movement either. Public opinion surveys at the time indicated that most residents (especially those in fast-growing urban areas) favored a strong federal role, which they saw as more likely to ensure access to valued recreational areas (Congressional Quarterly 1983). Without much political backing, the sagebrush rebels failed to overturn public policy.

The Reagan administration, and particularly James Watt's Interior Department, shared the rebels' political ideology and accommodated their demands to some extent (Culhane 1984). With the election of Bill Clinton, the same battles were joined once more under the new Wise Use label. BLM lands have been at the center of two of the most controversial proposals on Interior Secretary Bruce Babbitt's resource agenda: raising grazing fees and reforming the long outmoded 1872 Mining Law, which allows hard rock mining on public lands without any royalty payments and which contains few requirements for restoration of lands damaged by mining (Egan 1993a; Kriz 1993).

Babbitt's efforts build on a trend plainly evident in Congress before the 1992 election. For years, members had been seeking revisions in public lands and water policy to respond to increasing demands for preservation and recreation (Davis 1991). The shift could be seen in a diversity of measures, from the Omnibus Water Act of 1992 to the California Desert Protection Act of 1994. The former revised western water projects by instituting new pricing systems to encourage conservation and authorizing extensive wildlife and environmental protection, mitigation, and restoration programs; the latter designated some 7.5 million acres of wilderness on federal land within California.

Other Protected Lands and Agencies

Some of the public domain is spared the intensity of disputes that characterize decisions over use of Forest Service and BLM lands. In these cases, Congress has specified by statute that they be secure from much or all development once a decision is made to place the lands within a protected class. These are lands governed largely by the National Park Service and U.S. Fish and Wildlife Service (FWS). Box 5.1 lists the five major divisions of federal lands and one incorporating marine sanctuaries and estuaries.

The National Park Service   National Park System lands are closed to most economic uses, including mining, grazing, energy development, and timber harvesting, although such activities on adjacent land can seriously affect environmental quality within the parks (Freemuth 1991). These areas are managed by the National Park Service, an agency with 13,000 permanent employees (only about 120 of them scientists) and a budget of about $1.5 billion.

Several recent studies, including one from the National Academy of Sciences and another from the Commission on Research and Resource Management Policy in the National Park System sponsored by the National Parks and Conservation

Box 5.1 PRIMARY FEDERAL LAND SYSTEMS

National Park System

Created in 1872, the National Park System had grown by 1992 to contain 362 units, of which 50 were national parks, the "crown jewels" of the system. Most of those are in Alaska and the West. The NPS also includes over 300 national monuments, battlefields, memorials, historic sites, recreational areas, scenic parkways and trails, near-wilderness, seashores and lakeshores. The total covers over 80 million acres. Most NPS lands are closed to mining, timber harvesting, grazing, and other economic uses. Congress grants exemptions on a case-by-case basis when it approves the parks.

National Wildlife Refuge System

Created in 1903, the NWRS today contains 91 million acres, about 85 percent of which are in Alaska. They are the only federal lands specifically dedicated to wildlife preservation. There are 485 units in the system, including 49 national wildlife refuges. Within the NWRS are 18 million acres of forests on which limited commercial logging operations are permitted. Refuges are open to some commercial activities, including grazing, mining, and oil drilling. The system also contains waterfowl production areas, fish hatcheries, wildlife research stations, and other facilities.

National Forest System

Created in 1905, the NFS in 1991 included 187 million acres of land in 156 units, largely in the Far West, Southeast, and Alaska. There are 50 national forests in 23 eastern states as well. The NFS is managed chiefly by the Forest Service, although some BLM lands are included in the system as well. Recent controversies have focused on the degree to which forest lands should be preserved as wildlife habitat and for other ecological functions and the extent and location of commercial logging to be permitted.

National Wilderness Preservation System

Created in 1964, the NWPS in 1991 consisted of 95 million acres of land, 56 million of which are in Alaska and most of the rest in the 11 western states. The NWPS contains nearly 500 units of widely varying size. NWPS lands are set aside forever as undeveloped, roadless areas, without permanent improvements or human habitation. Only officially designated wilderness areas are protected from commercial development. Wilderness areas are not entirely separate from the other systems. They are found within the National Park System (over 38 million acres), National Wildlife Refuge System (over 20 million

Box 5.1   *Continued*

acres), and National Forest System (over 32 million acres), and on BLM lands. Congressional approval of the California Desert Protection Act in 1994 set aside some 7.5 million acres of protected wilderness, or about one-third of California's desert terrain.

National Wild and Scenic River System

In 1992 the river system contained 10,506 total miles in 151 rivers in one of three designations: wild, scenic, and recreational. Rivers are to be in free-flowing condition, that is, unblocked by a dam, and to possess remarkable scenic, recreational, ecological, and other values. The shorelines of designated rivers are protected from federally permitted development.

National Marine Sanctuaries and National Estuarine Research Reserves

In 1992 the marine sanctuaries consisted of 13 units and 9000 square nautical miles of unique marine recreational, ecological, historical, research, educational, and aesthetic resources, managed by the National Oceanic and Atmospheric Administration (NOAA). They are, in effect, equivalent to national parks in a marine environment. In 1992 the United States doubled areas protected as sanctuaries with four new units, including the Monterey Bay sanctuary, with over 4000 square nautical miles, second only to the Great Barrier Reef of Australia. In 1992 the estuarine research reserves consisted of 21 units and 420,000 acres protected by NOAA under the Coastal Zone Management Act.

National Rangelands

Located chiefly in the 11 western states and Alaska, rangelands consist of grasslands, prairie, deserts, and scrub forests. Much of the land is suitable for grazing of cattle, sheep, and other livestock. Over 400 million acres of range-land falls under federal control, most of it managed by the BLM and the rest by the Forest Service. Rangelands are the largest category of public domain lands.

Source: Compiled from Council on Environmental Quality, *Environmental Quality: 23rd Annual Report of the Council on Environmental Quality* (Washington: Government Printing Office, January 1993).

Association, have been highly critical of the National Park Service. The reports emphasize poor training of employees, eroding professionalism within the ranks, and a failure to educate the public on issues related to park lands. Criticism has been especially directed at the service's inadequate capabilities for scientific research that can be used in support of its resource protection policies (Kenworthy 1992b; Cahn 1993).

William Lowry's (1994) extensive analysis of the Park Service suggests that many of its problems can be traced to weakened political consensus and support for policy goals (especially long-term preservation of park lands) that have led to increased intervention by Congress and the White House to promote short-term and politically popular objectives. Budget woes persist and contribute to the problems. Visitors to the nation's parks increased some 25 percent between 1983 and 1992, but the Park Service budget remained about the same, adjusted for inflation.

Wilderness Areas and Wildlife Refuges   Like the National Park System lands, wilderness areas (primarily within the National Forest System, but including some BLM lands) are given permanent protection against development. This is less true, however, for wildlife refuges (governed by the Fish and Wildlife Service). In addition to preserving habitat for a diversity of wildlife, these lands are open to hunting, boating, grazing, mining, and oil drilling, among other private uses.

Environmentalists continue to argue for reducing those uses of wildlife refuges that are incompatible with protection of wetland, woodland, desert, and other fragile habitats. Supportive legislators have introduced such proposals in Congress. Under Bush's director of the Fish and Wildlife Service, John Turner, the agency began curtailing some incompatible human uses of wildlife refuges. It also embarked on a comprehensive inventory of refuges to identify those uses that could pose a threat to wildlife.

The Wetlands Dispute

One of the most prominent natural resource controversies of George Bush's presidency concerned one category of protected lands, those officially classified as wetlands. Of the estimated original 220 million acres of wetlands in the lower 48 states, only about 100 million remain. In the late 1980s, about 200,000 to 300,000 acres a year continued to be lost to development, with high rates of loss in Florida, Louisiana, Texas, Arkansas, Minnesota, and Illinois. Concern over the trend led Bush to announce in the 1988 election campaign that he would adopt a "no net loss of wetlands" policy in his administration.

Under the Clean Water Act's Section 404, the federal government regulates the dredging and filling of wetlands. Scientists in four agencies (the EPA, Army Corps of Engineers, the Department of Agriculture's Soil Conservation Service, and the Fish and Wildlife Service) had agreed on a new definition of wetlands consistent with present understanding of their ecological structure and functions (Alper 1992). It was to be incorporated into a technical manual to standardize and clarify what lands would be subject to restricted development. Different definitions were used prior to that time. The Bush administration agreed to back the new definition. Because it broadened the conventional view of wetlands, however, a National Wetlands Coalition, consisting of land developers, agricultural interests, and oil companies, tried to alter it and thus open more land to development. They appealed to the president's Council on Competitiveness, which intervened with the EPA, the lead agency, to demand a less restrictive definition. Under the approach the council favored, federal scientists estimated that between 30 and 80 percent of currently protected wetlands would be opened to commercial use.

The EPA received tens of thousands of formal public comments on the proposed rule change, with environmentalists hotly opposed. Scientific organizations uncharacteristically entered the fray as well, assisted by a National Academy of Sciences report recommending not only an end to loss of wetlands, but an ambitious program of wetland restoration to regain and preserve lost ecological functions (Stevens 1991). In 1992, Bush's EPA administrator, William Reilly, agreed under protest to accept a series of compromises that pushed the standard closer to what industry sought than agency scientists had originally proposed. Even after Bush's defeat in the November election, the competitiveness council continued to press for a looser definition, with speculation that the council's head, Vice President Dan Quayle, considered such a move to be beneficial to his expected run for the Republican presidential nomination in 1996 (Cushman 1992). The council backed off when Reilly refused to agree to the changes.

Six months into his presidency, Bill Clinton proposed broad new protection for wetlands. His recommendations provided for expanded coverage of wetlands in Alaska that Bush wanted to open to development. Clinton also opted to use the old 1987 definition of wetlands until the National Academy of Sciences completed a new study ordered by Congress (Johnston 1993). That authoritative report was issued in May 1995, and its endorsement of a broader definition of wetlands directly conflicted with an overhaul of the Clean Water Act being considered in the House. Despite the study's conclusions, members approved a bill that significantly weakened wetlands protection.

ENVIRONMENTAL IMPACTS AND NATURAL RESOURCES DECISION MAKING

Two of the most important environmental policies of the past 25 years merit special attention because of their wide application to public (and sometimes private) lands and their attempt to improve the way environmental science is used in natural resource decision making. These are the Endangered Species Act and the National Environmental Policy Act.

The Endangered Species Act

No other natural resource policy better captures the new environmental spirit of the past several decades than the Endangered Species Act (ESA). It is one of the strongest federal environmental laws, and it symbolizes the nation's commitment to environmental protection goals. For that reason, the ESA has become a lightening rod for anti-environmental rhetoric in the 1990s. The Fish and Wildlife Service, which implements the act for land-based species (the National Marine Fisheries Service handles water-based species), has been a target of frequent congressional intervention.

The goal of the 1973 act was "clear and unambiguous—the recovery of all species threatened with extinction" (Tobin 1990, 27). A species could be classified as endangered if it was "in danger of extinction throughout all or a significant portion of its range," or threatened if "likely to become an endangered species within the foreseeable future." Congress made the protective actions unqualified. All "takings" of such a species would be prohibited, whether on state, federal, or private land. Restrictions are, however, less stringent for plants than for animals; plants are not fully protected when they are on private land. In *TVA v. Hill* (1978), involving a tiny fish, the snail darter, the Supreme Court upheld the act's constitutionality and made clear that under the ESA, no federal action could jeopardize a species's existence—regardless of cost or consequences.

Congress modified the ESA several times after its initial enactment in attempts to mollify the critics. It even created a cabinet-level Endangered Species Committee, dubbed the God Squad, with authority to grant exemptions to the law.[9] As is often the case with environmental policies, the result is a set of complex and cumbersome procedures and decision rules: (1) determining whether a given species is either threatened or endangered and should be so "listed"; (2) designating critical habitats to be protected; (3) enforcing regulations that govern activities directly affecting the species and their habitats; and (4) implementing "recovery plans" for the species.

To complicate matters, Congress has never provided sufficient funds for the Fish and Wildlife Service to implement the act, and the agency has lacked the bureaucratic strength and resources necessary to fend off constituencies adversely affected by the ESA (Tobin 1990). Prior to transfer of some personnel to the new Biological Service, the FWS operated with about 7000 employees and a budget of slightly more than $1 billion a year.

Implementing the ESA: Achievements and Needs   The Council on Environmental Quality (1993) reported that as of December 1992, there were 755 U.S. species of plants and animals listed as threatened or endangered. The Threatened and Endangered Species List (which includes foreign species) had 1284 species; another 3964 were candidates for inclusion on the lists. After numerous critical studies and lawsuits over indifferent implementation of the ESA over the past decade, the government is under court order to accelerate the process of review for those candidate species. In late 1992 the Interior Department agreed to make decisions about the listing status of about 400 species on its top-priority list of candidates over the following four years.

To date, the FWS has made slow progress in designating critical habitats even for endangered species, and recovery of species on the list has been very limited. Many have become extinct while these bureaucratic processes drag on. One government study estimated that the nation would have to spend $460 million per year over ten years to develop recovery plans for all species that are candidates for listing. This is far in excess of the $60 million annual FWS budget for the program. However, the agency has publicized its relative success in protecting several of the "charismatic megafauna" that attract so much public attention, such as the American bald eagle and the California gray whale.

It must be said, however, that careful evaluations of the ESA do not support the harsh assessments of the act's detractors. These include Bush's interior secretary, Manuel Lujan, who called the ESA an impediment to economic progress. Some of the critics, including an alliance of timber, farming, ranching, and mining interests, have argued for radically changing the act to reduce its protection of species. Despite the publicity given to conflicts between protection of endangered species and federal projects such as logging on public lands, the act has prevented few projects from going forward, in part because adjustments are made to avoid disrupting critical habitats (Tobin 1990; Kenworthy 1992a; World Wildlife Fund 1994).

A greater problem with the act is its focus on individual species rather than the ecosystems of which they are a part. The northern spotted owl, for example, shares its habitat with over 1400 other species dependent on old growth forests; 40 of those are listed as threatened or endangered. Yet media coverage and political controversy focus on the owl, distorting public understanding of the real issues.

The National Biological Service   The ecosystem management approach favored by biologists and ecologists requires a greatly improved base of scientific knowledge. That is a major purpose of the Department of Interior's new bureau, the National Biological Service. Modeled after the U.S. Geological Survey, established in 1879 and now the nation's largest earth science research agency, the Biological Service is to consolidate the department's scientific research in an effort to inventory and monitor all plant and animal species in the nation and their habitats. About 1600 Interior employees will be transferred to the service, including 1200 scientists.

Secretary Babbitt argues that such a comprehensive survey of biodiversity would head off the kinds of battles represented by the spotted owl, what he calls "train wrecks," or collisions of principles in which each side loses. The survey would make possible a proactive approach that could prevent species from becoming endangered in the first place by using the knowledge base for appropriate balancing of conservation and development.

Congressional critics of the survey, led by Rep. W. J. "Billy" Tauzin (D-Louisiana), have argued that it could pose a threat to private property rights if owners are prevented from developing their property. Their objections to the legislation temporarily prevented formal authorization of the survey by Congress, although the activity has been funded by appropriation committees at about $167 million a year. Members approved amendments to these bills that require written permission from landowners before any surveys of their property can take place. Babbitt is using his administrative authority to go ahead with the biological inventory (Camia 1994c).

Renewing the ESA and Resolving Conflicts   In an effort to appease opponents who have blocked renewal of the Endangered Species Act, the Clinton White House in June 1994 proposed administrative changes intended to ensure that species recovery plans are "scientifically sound and sensitive to human needs." New scientific peer review processes would be used, representation on planning bodies would be broadened, private landowners would be apprised early in the process about allowable actions, and multispecies listings and recovery plans would receive emphasis (New York Times 1994b).

In much the same accommodationist vein, the Clinton administration also sought to end the long conflict over protection of old growth forests and the spotted owl by convening an extraordinary Forest Conference in 1993. The one-day conference, with a third of the president's cabinet in attendance, included participation by loggers, environmentalists, scientists, salmon fishers, Native Americans, mill workers, small business owners, and timber industry executives, and it sought to find common ground. In the end, Clinton proposed a compromise that would set aside much larger areas of the forest than under earlier government plans to protect the owl and other species—and the area's watersheds and streams. It would, however, protect fewer lumber worker jobs. The government estimated that 9500 timber-related jobs would be lost under its plan. However, Clinton proposed funding job retraining and creating other economic revitalization programs for forest-dependent communities.

Previous attempts by Congress and the White House had failed to produce an acceptable solution to these conflicts in the Pacific Northwest, or they were declared invalid by the federal courts because they were inconsistent with environmental law. In June 1994 Judge William L. Dwyer of the Federal District Court in Seattle, who had long overseen the spotted owl case, lifted a three-year ban on logging on Forest Service lands in the region pending the outcome of the latest round of lawsuits and his own review of the plan. Both environmental and industry groups had called the Clinton forest plan unacceptable and had filed legal challenges to it.[10] In December 1994 Judge Dwyer approved the plan, giving the Clinton administration a rare, though only temporary, victory in environmental policy.

International Efforts to Protect Species   International agreements, especially the Convention on International Trade in Endangered Species of Wild Fauna and Flora (CITES), also play a role in protection of species. U.S. leadership led to approval of CITES, which became effective in 1975. The agreement is overseen by the U.N. Environment Programme in cooperation with the nongovernmental International Union for the Conservation of Nature (IUCN) and the World Wildlife Fund (Caldwell 1990). Other private groups, such as the Nature Conservancy, contribute through negotiating land donations or outright purchase of habitat. CITES is supposed to operate through permits that restrict or prohibit international trade of species in different categories of protection (those threatened with extinction, those that may be threatened, or those whose exploitation should be prevented). However, the agreement has yet to end an international market in animal smuggling, especially in Southeast Asia. Japan has been a particular target of environmentalist protest, and African nations have regularly argued with decisions related to protection of the elephant (Switzer 1994).

The National Environmental Policy Act

Few people in 1969 anticipated the impacts the National Environmental Policy Act would have on decision making across the entire federal government. Over the past 25 years, this "procedural" policy has transformed expectations for the way government agencies should consider the effects of their actions on the environment. Its influence on land-use decisions affecting natural resources has been especially striking. NEPA's success owes much to the entrepreneurial use of the environmental impact statement process by environmentalists and administrative leaders to advance environmental values, both at the federal level and in parallel cases at the state and local level where similar impact assessments under "little NEPAs" are required (Caldwell 1982; Bartlett 1989). Some states, such as California and Washington, have been particularly demanding in their requirements for impact statements for both state and private projects.

Even with later amendments, NEPA remains a brief statute at six pages (Council on Environmental Quality 1993, 444–49). Section 101(a) acknowledges the "profound impact of man's activity on the interrelations of all components of the natural environment" and the "critical importance of restoring and maintaining environmental quality to the overall welfare and development of man." The instrument for achieving those goals is the EIS process, which was to use "a systematic, interdisciplinary approach" to ensure the "integrated use of the natural and social sciences" in planning and decision making. As stated in section 102(2)(c), EISs are to offer a detailed statement on the environmental impact of the proposed action, any adverse environmental effects that cannot be avoided, alternatives to the action contemplated, the relationship between "local short-term uses" of the environment and the "maintenance and enhancement of long-term productivity," and any "irreversible and irretrievable commitments of resources" which the proposed action would require.

In addition, the statute calls for wide consultation with federal agencies and publication of the EIS for public review. The intent was not to block development projects, but to open and broaden the decision-making process. As was the case with protection of species under the ESA, few actions have been halted entirely. Instead, agencies follow one of two courses of action. The first is that they no longer even propose projects and programs that may have unacceptable impacts on the environment. The second is that when they do both propose and move ahead with projects and programs, they employ mitigation measures to eliminate or greatly reduce the environmental impacts (Council on Environmental Quality 1993, 153).

During the 1970s and 1980s, federal agencies grew accustomed to NEPA requirements and to public involvement in their decision making. Some, like the much-criticized Army Corps of Engineers, dramatically altered their behavior in the process (Mazmanian and Nienaber 1979). Others, such as the Department of Energy, made far less progress in adapting to the new norms of open and environmentally sensitive decision making. They serve as a reminder that statutes alone cannot bring about organizational change (Clary and Kraft 1989).

The CEQ has tried to help with the process of organizational adaptation. It is responsible under NEPA for supervising the EIS process, and it has worked closely with federal agencies through workshops and consultations to define their NEPA responsibilities. With a series of court rulings, CEQ regulations have also clarified the extent of the mandated EISs and the format to be used to enhance their utility. CEQ regulations distinguish Environmental Assessments (EAs) from EISs. The EAs are more limited in scope and more concise, and are used where a project and its impacts do not require a full EIS under NEPA. Agencies also keep a Record of Decision (ROD), a public document that reflects the final decision, the rationale behind it, and commitments to monitoring and mitigation.

As noted above, the EIS process has been used frequently by environmental groups to contest actions of natural resource agencies, such as the Forest Service and the Bureau of Land Management. The pattern continues. In 1992, for example, the total number of draft, final, and supplemental EISs filed by all federal agencies totaled 512 (Council on Environmental Quality 1993, 162). The greatest number in recent years have come from the natural resource agencies, with the most (in order) from the Agriculture, Transportation, and Interior departments, and from the Army Corps of Engineers. There are many more EAs than EISs prepared each year. For

1992, federal agencies prepared some 45,000 EAs, not all of which are as concise as might be expected. Some of the EAs submitted in the 1980s for the Energy department's nuclear waste disposal site evaluations, for example, ran to over 1000 pages.

Unfortunately, many impact statements still fall short of the expectations for comprehensive assessments of likely environmental impacts and thus disappoint those who hoped that the mandated administrative process would "make bureaucracies think" about the consequences of their actions. Some impacts have been ignored as unimportant and others have proved hard to forecast accurately. Obviously, limited knowledge of biological, geological, and other natural systems constrains anyone's ability to forecast all significant impacts on the environment of development projects. Thus improvement of EISs will require expanded knowledge of ecosystems as well as of related natural and social systems. For many projects, policy makers of necessity have to make decisions under conditions of some uncertainty.[11]

Partly because of these deficiencies, court cases under NEPA continue. In 1991, 94 cases were filed where there was a NEPA claim against the federal government. The Department of Transportation was the most frequent defendant, with the Interior Department second, and the Army Corps third. The most common complaint has been that no EIS was prepared when one should have been or that the EIS was inadequate. The most frequent plaintiffs by far in these cases have been individuals and citizen and environmental groups, as has been the case since 1970. The number of cases each year has varied, from a high of 189 in 1974 to a low of 17 in 1982, with an average for the 1974 to 1991 period of 104 per year (Council on Environmental Quality 1993).

THE CHANGING NATURAL RESOURCES POLICY AGENDA

A review of natural resource policies and agencies conveys much about dominant patterns of resource use in the nineteenth and twentieth centuries—and their consequences. It also indicates that the challenges mounted by the environmental community to such resource use over the past three decades have been reasonably successful. The American public and its policy makers have broadly endorsed conservation goals. Congress has provided most of the statutory authority needed to halt environmentally destructive practices and to move toward a future course of sustainable development. However, it also has continued to support western extractive industries and it has not completely reconciled inherent conflicts with goals of environmental sustainability. Nor have recent presidents. As George Miller, chair of the House Natural Resources Committee, stated in 1993, "Reagan and Bush were just holding back the future [on resource management]. . . . They were the last gasp of an outdated philosophy" (Egan 1993a). He might have said much the same of many of his colleagues in the House and Senate and of previous Democratic presidents.

The Clinton administration began its tenure with a strong and much ballyhooed commitment to a new natural resource agenda. Secretary Babbitt announced his intention to roll back a century of ill-advised resource practices encouraged by government subsidies and lax enforcement of environmental laws. "It's a brand new era in land management," Babbitt said (Egan 1993a). He sought to end "below-market" sales of timber from national forests, charge royalties for the first time on gold, silver and other hard rock mining on public lands, and raise the fees that ranchers pay to graze livestock on nearly 300 million acres of public rangeland, which have been well below prevailing rates charged on comparable private land.[12] The changes had long been called for by economists, policy analysts, land managers, and biologists, but they would have to run a difficult political gauntlet to survive. They didn't make it through unscathed.

Some of Babbitt's proposals, such as revision of the 1872 Mining Law, required congressional approval. For others, such as higher grazing fees, such a move was merely politically advisable. That approval was not forthcoming either on the grazing fees or on reform of the mining act. Despite Babbitt's extended efforts to design an acceptable compromise that included these policy changes, in 1993 and again in 1994, western senators successfully filibustered the proposals. The National Cattlemen's Association and the American Mining Congress, both major campaign contributors and vociferously opposed to the reforms, assisted in blocking them. However, Congress did approve a one-year moratorium on new mining patent applications to halt the exploitation of mineral rich lands; patents allow miners to buy title to federal land for as little as $2.50 an acre. Representative Miller promised to fight for a permanent ban until the mining law itself could be reformed, though his influence declined sharply in the new Republican Congress.

The press called the opposition to Babbitt's resource proposals an indication of a second Sagebrush Rebellion brewing in the nation. The protest was actively cultivated by the Wise Use movement and was tied to a broader backlash against environmental policy both in Congress and at the state and local level. Senator Ben Nighthorse Campbell (D-Colorado), a leader of the opposition to Babbitt's proposals, and other western senators spoke of an alleged "war on the West" being waged by environmentalists and the Clinton administration. Campbell said in one interview, for example, that "these land-use reforms look like an all-out assault on the West and how people earn a living in the West" (Kriz 1993).[13]

Most of the western senators echoed the hyperbole that the resource policy changes would mean economic devastation for ranchers and farmers in the West. Senator Alan Simpson (R-Wyoming), saying he was defending a cherished western lifestyle, argued that the higher fees would "do those old cowboys in." Following a review of the top 500 holders of grazing permits on federal land, the *New York Times* reported that the cowboys in this mythical West were more likely to wear wingtips than boots. They included some of the wealthiest families in the nation, the Metropolitan Life Insurance Company, a Japanese conglomerate, the Mormon Church, and the Nature Conservancy. Only 28,000 people out of the 50 million who live in the 11 western states hold grazing permits, and the smaller holders have little in common with the largest (Egan 1993b).

It is true that many ranchers operate on a very slim profit margin, which could be jeopardized by a substantial or rapid increase in grazing fees. Yet environmentalists argue that alone is no reason to continue to subsidize activities that lead to degradation of fragile rangelands. Studies by the General Accounting Office and private groups confirm that current grazing practices risk long-term environmental damage (McInnis 1993). Despite these concerns, the Simpson forces won on grazing fee reforms in 1993, and President Clinton himself withdrew from the battle (Berke 1993).

Babbitt vowed to pursue most of the policy changes through the administrative rule-making process. He also was anxious to build more backing for his plan in the West. In early 1994 he spent several months in the western states trying to sell a compromise measure that would provide local interests—including environmentalists and state and local officials as well as ranchers, miners, and timber industry representatives—with more influence over public land decisions through regional and advisory councils designed to build consensus. The federal government would set general goals for environmental protection, but states and the advisory councils would set specific standards for rangeland use. During these intense negotiations, Jim Baca, the Clinton-appointed director of the BLM, resigned over policy differences and management style, further eroding Clinton's reputation among environmentalists. Later in the year, Babbitt held public hearings in the West on his proposed administrative regulations, incorporating the revised grazing fee policy as part of his continuing effort to secure approval for it. Babbitt argued that up to 80 percent of the people living in the western states supported reform of the grazing and mining laws. Yet by early 1995 he retreated from the proposed reforms and pledged to work cooperatively with the new Republican Congress, some of whose members favored drastic cuts in the Interior Department's budget and programs.

CONCLUSION

Much like environmental protection policy, criticisms of energy and natural resources policy abound. They come in a variety of forms, as do the proposals for reform, and many of them were noted throughout the chapter. Some are similar to those associated with environmental protection policy, for example that regulations be more flexible to take into account local conditions and that stakeholders be given sufficient opportunity to voice their concerns. Also similar, but with environmentalists making the case this time, is that market incentives be used to improve environmental protection, particularly that user fees be set at a level that discourages environmental degradation and fosters sustainability. Improved scientific knowledge and better integration of decision making across programs and agencies are common suggestions here as well, and they are crucial to developing the institutional capacity for ecosystem management and sustainable development.

As repeated efforts to devise acceptable and legal solutions to the protection of old growth forest ecosystems in the Pacific Northwest attest, natural resources policy also requires the development of creative ways to involve the American public in policy decisions which affect their communities and livelihoods while maintaining adherence to professional norms of natural resource management. Some would dismiss such hopes as unrealistic. Yet experience in the Pacific Northwest and elsewhere suggests that the economic livelihood of rural communities can indeed be reconciled with resource conservation. It has been achieved through small-scale efforts to develop environmentally sound businesses and through the realization that the future economy of many communities depends more on maintenance of recreation and tourism opportunities than on traditional extractive industries. At this level, removed from the ideological national debates over balancing the economy and the environment, new approaches seem to work. For example, communities have used "facilitated dialogue" to develop a "common understanding of a problem and seek mutually beneficial solutions" (Johnson 1993). Like other forms of environmental mediation, such a dialogue seeks to build trust, broaden the range of participants, and educate people on the complexities of the issues.

These experiments in environmental mediation and consensus building are promising though they will be harder to apply on a larger scale. Such approaches face particularly difficult obstacles when participants hold passionate and conflicting views about the values at stake—protection of ecologically critical and aesthetically treasured wilderness areas; access to valuable timber, energy, water, and mineral sources; and protection of property rights and jobs in economic hard times. Many people will see little reason to compromise on what they consider fundamental principles.

In this vein, one of the most striking developments in recent years has been the proliferation of "property rights" bills in state legislatures following the Supreme Court's 1992 *Lucas* case. Such bills commonly demanded one of two actions: (1) governmental compensation to landowners for environmental restrictions (e.g., on use of wetlands) in proportion to the percentage of their investment lost, or (2) some kind of mandatory assessment of the impact of proposed agency action on property rights. Carried to their extreme, either kind of measure could well render governments powerless to regulate environmental quality. By mid-1994, bills incorporating property rights provisions had been introduced in 40 state legislatures and passed in 11. Most of those enacted were heavily amended to address environmental concerns. Environmental activists singled out measures in three states—Arizona, Mississippi, and Utah—as potentially serious threats to environmental protection. Ironically, the property rights and Wise Use movements modeled this legislation after NEPA's environmental statement requirement.[14]

By some accounts the property rights movement was picking up momentum in early 1995 (Schneider 1995). Yet in one crucial test in November 1994, 60 percent of Arizona voters rejected Proposition 300, a ballot measure that would have activated the nation's first state-level "takings" law. If approved, the proposition would have required the state to consider whether any proposed environmental regulation violated a citizen's property rights before putting it into effect. The measure's defeat meant that the law approved by the state's legislature in 1992 to limit takings of private property would not go into effect.

As is increasingly evident, many environmental issues *cannot* be brokered in the way policy makers are accustomed to treating other policy disputes. We have been slow to realize the distinctive problems some environmental issues present, and as a result we keep trying to force them into conventional categories. For many policy makers, it has been business as usual as they fight to protect influential constituencies no matter what cost to national policy objectives. Given the persistence of such practices, to make sustainable development a reality in the 1990s and the twenty-first century requires that we search for new ways to make these critical choices about our collective environmental future. This is as vital a need in the many arenas outside government that shape energy and natural resources as it is within govern-ment itself. We should also recognize the very considerable progress that has been made since the early 1960s in the nation's energy and natural resources policies. Much has been accomplished even if a great deal remains to be done.

ENDNOTES

1. The Wise Use movement emerged from a Multiple Use Strategy Conference held in 1988. It was convened by the Center for the Defense of Free Enterprise, a conservative educational foundation in Bellevue, Washington, that continues to serve as the move-ment's philosophical nucleus. Some accounts put the groups' collective membership at about 16,000 people, although they use a mailing list of 1.4 million to preach their gospel. The three most visible leaders of the movement are Charles Cushman, Ron Arnold, and conservative fund-raiser Alan Gottlieb. For further detail, see Poole (1992), O'Callaghan (1992), and Bosso (1994). Camia (1994c) reports on the impact of both the Wise Use and related property rights movement on environmental policy debate in Con-gress.

2. DOE calculates the subsidies differently and finds only a $5 billion to $13 billion annual subsidy, which it considers negligible given the size of the nation's energy bill (Wald 1992).

3. A portion of federal, state, and local taxes on gasoline supports road construction and maintenance, but it covers only about 60 percent of the cost. The rest is subsidized with general tax revenues. About 40 cents of the 1994 price of $1.15 per gallon is tax. In most of Western Europe, taxes are far higher, raising the total price of gasoline to about $3.50 a gallon.

4. These figures reflect my own calculations of budgetary change in these periods, drawn from the annual federal budgets and adjusted for inflation.

5. The oil drilling case is drawn from a much broader review of environmental policy in the Bush administration (Kraft and Vig 1990).

6. Bush differed less from Reagan on natural resource issues than he did on environmental protection. His appointees to natural resource agencies were similar to Reagan's. For his secretary of interior, Bush named Lujan, who had a generally poor environmental record in his 20 years in the House of Representatives. He was regarded as a strong backer of western mining, oil, and timber interests, and his LCV score averaged only 23 percent between 1970 and 1988.

7. For a fuller description of natural resources policy, see Henning and Mangun (1989). Al-though somewhat dated, a highly readable account of natural resource programs, partic-ularly in the Reagan years, is Congressional Quarterly's *The Battle for Natural Re-sources* (1983).

8. Federal authority to regulate land use comes from the Constitution, Article 4, Section 3, which gave Congress the "power to dispose of and make all needful rules and regulations respecting the territory or other property belonging to the United States."

9. As part of the spotted owl controversy, BLM director D. Cy Jamison petitioned the Sec-retary of the Interior, Manuel Lujan, to convene the Endangered Species Committee to consider 44 BLM timber sales in Oregon that FWS said would jeopardize the spotted owl. In May 1992 the committee voted to suspend the act and to allow logging on 13 of the 44 proposed sites, only the second time in history that the committee had voted to override the ESA. The decision was later overturned by the courts (Council on Environ-

mental Quality 1993, 118–19). Environmentalists had sued to block the exemptions on the grounds that the Administrative Procedure Act was violated when Bush administration officials summoned committee members to the White House and pressured them in an off-the-record meeting to grant the exemptions. A federal appeals court agreed that such communications are illegal.

10. For an extended treatment of the prominent spotted owl case, see Steven Lewis Yaffee's *The Wisdom of the Spotted Owl* (1994).

11. For a review of NEPA's application and assessments of its effects on environmental decision making, see a recent symposium edited by Bartlett and Malone (1993).

12. As part of its broader pledge to impose user fees, the administration also endorsed, as did environmental groups, higher fees for individual recreational use of public lands such as camping and boating.

13. As a further sign of his disaffection, in 1995 Campbell announced that he was leaving the Democratic party and joining the Republicans.

14. A Supreme Court case decided in June 1994 on a 5-4 split vote, *Dolan v. City of Tigard,* was applauded by property rights advocates as supporting their crusade, though it is likely to be fairly limited in its impact. The Court held that requiring public easements (e.g., green spaces) as a condition for a building permit was an unconstitutional "taking" of property unless government can show a "rough proportionality" between the requirement and the harm posed by the development. The Court also put the burden of proof on the government to justify its land-use restrictions, a reversal of previous law.

CHAPTER SIX

Evaluating Environmental Policy

Demands for reinventing government have become as commonplace in the mid-1990s as arguments for reengineering the corporation. In the public sector, such an enterprise should aim to make government better able to recognize and respond to pressing problems, and to do so with the policy approaches that promise the most effective solutions. If the need for reinvention is measured by the frequency or severity of critical assessments, there are few areas more deserving of revitalization than environmental policy.

As we saw in Chapters 4 and 5, business and industry groups complain that the nation overregulates in an irrational effort to eliminate or reduce minuscule risks to public health and the environment and thereby imposes unnecessary burdens on individual firms and the economy as a whole. State and local governments struggling to meet federal environmental mandates insist on greater flexibility and more federal funds to cope with their manifold responsibilities. Natural resource users such as ranchers, loggers, miners, and farmers fight to prevent what they view as precipitous and unwarranted loss of long-held federal subsidies that have helped to assure them of financial success. Workers in those industries and the communities in which they live often blame environmental policies for threatening their economic livelihood.

Analysts at conservative think tanks such as the Heritage Foundation, Pacific Research Institute, and Cato Institute object in principle to regulation and natural resource decision making dominated by the federal government. As an alternative, they praise the shifting of environmental responsibilities to the state and local level and reliance on private markets (Anderson and Leal 1991; Greve and Smith 1992).[1] They are joined in many of those positions by the property rights and Wise Use movements.

Environmentalists are unhappy as well, but for different reasons. They applaud the strong policies adopted over the past several decades, but they argue that too frequently environmental protection measures are compromised and unenforced by public officials insufficiently committed to their goals. They believe that too much regulatory flexibility invites weak enforcement and slow progress. They also fear that "consensus decision making" on natural resource issues may tilt excessively toward development at the expense of ecosystem health. Grassroots citizens groups see too few achievements that improve the quality of their local environments, and they worry about the continuing impacts on their health of air and water pollution and toxic chemicals. They want a greater say in cleanup plans and the location and operation of new industrial facilities.

Some of these criticisms have more merit than others, but all must be taken seriously. Over the past 25 years, both government and industry have invested a vast sum of money in scientific research, technological development, and pollution control and abatement. Total *federal* spending on environmental and natural resource policies runs at $22 billion a year (fiscal 1994), or about 1.5 percent of the federal budget of $1.5 trillion. Yet state and local governments and the private sector pay most of the costs of federal environmental protection policies. That cost is certain to rise over the next decade. It is reasonable to ask what such expenditures—and all the other policy efforts—bring the public in return. Are the costs justifiable in light of the goals of environmental policies, achievements to date, and the promise of future improvements in environmental quality?

Using these and other criteria, we need to ask which policies and programs have proven to be the most effective and which the least, and why. What impacts have they had, intended and unintended, and what are the implications for policy redesign? How might policy analysis and program evaluation help in determining which approaches to environmental policy make the most sense in particular situations, and in making changes where appropriate?

This chapter focuses primarily on U.S. environmental protection and natural resource policies. I try to respond to some of the most important critiques by examining selected evidence of program accomplishments and other impacts and considering alternative approaches to environmental policy. Chapters 4 and 5 highlighted deficiencies of particular policies; here I emphasize the broader arguments that cut across policy areas.

APPROACHES TO ENVIRONMENTAL POLICY EVALUATION

Policy and program evaluations come in many forms and differ in their scope of inquiry and analytic rigor. They include investigations by congressional committees and other arms of Congress such as the General Accounting Office and the Office of Technology Assessment; studies by industry, environmental groups, think tanks, and academic institutions; and internal assessments conducted by the agencies and departments themselves. Given the variable standards applied in such evaluations, policies and programs judged failures from one perspective may be considered ringing successes from another. Consumers of environmental policy evaluations would be well advised to buy and use them with caution.

In theory, evaluation and reform of environmental policies and programs is straightforward. One should simply ask whether the stated goals of environmental statutes have been achieved or not, and perhaps at what cost. The answers should guide policy and program changes. In reality, assessing environmental policy and improving its effectiveness is much more complex and intertwined with the political process. There are winners and losers in policy choices, and each side will take an active interest in evaluations that may maintain, alter, or eliminate programs affecting them. Supporters may shield some programs from critical assessment while those that are politically vulnerable may have repeated evaluations thrust upon them by well-placed critics.

None of these constraints diminishes the genuine need for environmental policy and program evaluations (Knaap, Kim, and Fitipaldi 1996). Their value lies in contributing to systematic, critical, and independent thinking in the policy-making process, for which there is a rising demand in the 1990s. For this reason, evaluations are likely to become more common in the future.

For students of public policy, evaluation means appraising the merit of governmental processes and programs. The term *policy evaluation* usually refers to the assessment of alternative policy proposals. *Program evaluation* means judging the success of programs already in place and, especially, determining whether and how they affect the problems to which they are directed. We look for program impacts on individual behavior, institutional processes, and eventually environmental quality itself. Most environmental evaluations understandably focus on program outcomes. Yet those assessments must be supplemented by some consideration of what is called process and institutional evaluation (Bartlett 1994).

Program Outcomes

In outcomes evaluation, analysts compare measures of environmental quality outcomes with policy objectives. In effect, we are asking whether the air and water are cleaner, the drinking water safer, the hazardous waste sites cleaned up, and endangered species protected. I reviewed data of this kind in Chapter 2. Both proponents and critics of environmental programs use outcome measures to demonstrate either impressive policy achievements or serious shortcomings, with implications for improved implementation, redirection of program priorities, or policy change of a more extensive form.

Yet questions about program outcomes are difficult to answer definitively, and evaluations of them suffer from at least six major limitations that must be borne in mind in reviewing such evidence.

1. It is difficult to select the outcomes to measure when Congress fails to specify clear policy goals and objectives and realistic timetables for achieving them. The same is true for policies that mix substantive and symbolic goals, as many environmental policies do.
2. The quantity and quality of available data often are inadequate, especially for judging long-term trends. In many cases, there may be so little valid data (as a result of insufficient or uneven monitoring of environmental conditions) that no meaningful conclusions about achievement of program goals and objectives are possible (Russell 1990). We simply don't know as much as we should about whether and in what ways environmental quality is changing.

3. Sometimes the indicators used are measures not of environmental outcomes at all but rather of bureaucratic tasks (e.g., the number of pollution permits issued or enforcement actions taken). The GAO refers to this as using "activity-based indicators" (critics call it bean counting), and it has faulted the EPA for relying on such measures in tracking progress in pollution control (U.S. General Accounting Office 1993d). We must distinguish such indicators from actual measures of change in environmental quality, public health, and ecosystem health, which constitute the real "bottom line" of environmental policy.

4. The effectiveness of environmental programs cannot be judged merely by looking at outcomes and without regard to causality. Some environmental improvements may be due to factors other than program activities, such as technological progress and actions industry would have taken anyway as part of its modernization programs and to compete more effectively in the marketplace (Portney 1990b).

5. National data on policy or program results aggregate and may mask what is often substantial variation among states and localities. Findings that show national improvement, or lack of progress, may or may not accurately characterize individual states, regions, or communities. They may be making greater or less progress than the nation as a whole.

6. Studies that examine outcomes alone cannot address important questions about program operations. If such studies find program achievements to be modest to nonexistent, one may not learn *why* that is the case, and thus be left with no guide to appropriate policy and program changes.

Process and Institutional Evaluation

Because of these limitations, outcomes evaluation in some cases needs to be supplemented by process and institutional evaluation. Process evaluations attempt to evaluate the merit of decision-making processes themselves. We might ask, for example, how well those processes measure up to the standards of comprehensiveness, promotion of ecological rationality, or integrated decision-making, which are crucial to effective environmental policy and yet hard to achieve in practice (Dryzek 1987; Bartlett 1990). If such decision-making capacities are found lacking in government agencies, how might we improve the situation? Where new approaches such as ecosystem management or integrated permitting for new facilities are adopted, we should ask how they affect decision making and whether they produce the results expected.

We might ask as well whether environmental policy processes have been responsive to public needs or incorporate sufficient participation by significant interests. Those questions lie at the heart of improving hazardous waste programs that require community participation in cleanup decisions. Much of the argument about environmental injustice in the treatment of poor and minority communities concerns such process standards (Bullard 1993). Critics raise similar issues in evaluation of nuclear waste disposal policies. Study after study has documented distrust of the Department of Energy, public perceptions of risks at odds with departmental assurances, and the ineffectiveness to date of its approaches to citizen involvement in decisions on siting of disposal facilities (Dunlap, Kraft, and Rosa 1993; Slovic 1993).

As these examples indicate, process evaluations speak to issues disregarded by outcomes evaluation. They are not directed at substantive changes in environmental quality, nor are they preoccupied with the standards of efficiency that so concern economists. Rather they are intended to determine the success and failure of public policies along other dimensions. Some of those dimensions contribute to our understanding of why we get the results we do and thus speak to the question of what might be done by policy activists, legislators, and agency personnel to redesign policies or improve programs.

Institutional evaluations cast an even broader net. They ask how environmental policies affect institutions, such as bureaucratic agencies—and also how institutional characteristics shape the kind of environmental policies we get and their likely success. For example, over a decade or more, environmental policies may shape bureaucratic culture, including prevailing beliefs, values, procedures, and incentives. There is abundant evidence that the National Environmental Policy Act had such effects as federal agencies over time adapted to the requirements that they assess environmental impacts of their decisions (Caldwell 1982; Bartlett 1989). Much the same seems to be occurring with hazardous waste policies. They have altered our perception of the acceptability of hazardous wastes and have increased costs of treatment and disposal. Both should lead to significant reductions in the quantity of hazardous waste generated, in part through new manufacturing processes.

The effects of institutional characteristics are evident on both Capitol Hill and in the executive branch. Barry Rabe (1990), for example, argues convincingly that dispersal of power in Congress is responsible for many of the shortcomings in the Superfund program, such as statutory inconsistencies, lack of priority setting, and overestimates of the technical and administrative capacity of the EPA to implement the program. In the executive branch, the convoluted, reactive, and fragmented process of environmental rule making is a source of endless complaints from both business and environmentalists. The issue was directly addressed in the final report of the Carnegie Commission on Science, Technology, and Government (1993a). The commission called for improved coordination and strategic planning for environmental regulation, and urged reform of the Executive Office to allow it "to reach out across a fragmented government to begin reformulating policies toward a more sustainable future" (57). Vice President Al Gore's National Performance Review (1993b) made much the same case for instituting an interagency regulatory coordinating body to bring greater coherence and efficiency to environmental and other regulation.

Despite these important qualifications about what program evaluations can and cannot reveal about the success or failure of environmental policies, some conclusions are possible. The key questions remain the extent to which environmental policy achieves its goals and objectives, and at what cost and with what other important effects. For this purpose, I combine environmental protection, energy, and natural resources policies, noting important differences where useful.

SIGNS OF PROGRESS

The data on outcome measures presented in Chapter 2, and to a lesser extent in Chapters 4 and 5, provide good evidence of progress in meeting environmental quality goals in some areas while falling short in others. As the earlier review indicated, no simple generalization can capture the full story across all environmental protection and natural resource policies. Yet a strong case can be made that conditions would likely be substantially worse today if the major environmental policies were not in place.

Environmental Protection Policies

Such a conclusion is particularly valid for air and water pollution control where enormous gains have been recorded since the early 1970s. Even better results might have been obtained in these and other areas had sufficient resources been provided to the EPA and state agencies, had the programs been better managed, had the federal government established better working relations with the states, and had less time been spent by all parties in drawn out administrative and legal proceedings.

Where additional improvement must be made, for example, in controlling toxic chemicals, cleaning up abandoned waste sites, and reducing hazardous air pollutants, these conclusions should inspire at least some confidence in the regulatory approaches that are so widely disparaged in the 1990s. They should also help to suggest the kinds of policy and administrative changes that might improve the effectiveness and efficiency of environmental policies and yield better results in the future. By any measure, for example, hazardous waste remediation will preoccupy government and industry for decades to come. A very large number of sites need cleanup and the nation will spend hundreds of billions of dollars over the next 30 to 50 years in pursuit of that goal. The scope of these activities demands that programs be well designed and managed and that the funds be used efficiently. Yet evaluations by the GAO, the OTA, and independent analysts suggest that current remediation efforts fall short on all accounts. Clearly, we need to do a better job (Kraft 1994c).

Perhaps the greatest environmental policy success has been recorded in air pollution control. Several different indicators show that emissions of pollutants have declined impressively and air quality is improving nationwide. The latest reports on air quality from the EPA confirm the long-term trend despite continued population and economic growth and increased reliance on the automobile (U.S. Environmental Protection Agency 1993). At least some of the improvement is clearly attributable to enforcement of the Clean Air Act.

The most persuasive evidence on that score is at the state level. Ringquist's study comparing the 50 states shows that "strong air quality programs result in decreased levels of pollutant emissions" even when controlling for other variables such as the states' economy and politics. The stronger programs produce greater reductions in ambient air pollutants. Enforcement is a key factor. States that vigorously enforce controls on stationary sources have lower emissions. The most important variables are consistency in enforcement and well-focused and well-supported administrative efforts (Ringquist 1993, 150–51). The states would be less effective in producing such outcomes without a powerful federal EPA to back them up as a "gorilla in the closet," and thus spur enforcement actions that regulated parties will take seriously.

Some of these same conclusions apply to water pollution control. The nation's water quality has improved significantly over the past 20 years thanks to the Clean Water Act and the Safe Drinking Water Act. However, progress here has been much more uneven and slower in coming than in air quality. Especially notable have been substantial reductions in discharge of pollutants from point sources and advances in drinking water quality, especially in cities. In contrast, controlling nonpoint sources has enjoyed only minimal success and remains a major focus of current water quality program efforts. Groundwater quality in many areas continues to deteriorate as well.

The picture is similarly mixed on toxic chemicals and cleanup of hazardous waste sites. The annual Toxics Release Inventory reports show important reductions in releases of toxic chemicals by major industrial sources (thanks to mandatory public disclosure of emission data). Cleanup actions under Superfund have accomplished far more than usually acknowledged (De Saillan 1993). The EPA's 33/50 initiative called for voluntary efforts by industry to reduce the use of 17 priority toxic chemicals below the original 1988 TRI levels—33 percent by 1992 and 50 percent by 1995. The agency reports impressive progress toward those goals, and companies like Mobil Oil proudly advertise their achievements. New EPA rules promise to reduce the amount of hazardous chemicals stored at thousands of industrial facilities nationwide by forcing companies to disclose the amounts of such chemicals stored on the premises and their health risks (Holusha 1994). Yet some citizen groups have sharply criticized these voluntary efforts, and especially the 33/50 program, as largely a public relations ploy. They prefer a strong national policy on pollution prevention.

In all these cases, deficiencies in the major environmental protection policies have not gone unnoticed. In revising these acts, Congress has had at its disposal abundant critical analyses and proposals for policy change even if its final decisions represent compromises among conflicting interests that are unlikely to optimize environmental policy.

Natural Resource Policies

Judging the success of natural resource policies is no easier than determining whether clean air and water policies are working. The kinds of measures used here, acres of "protected lands" set aside in national parks and wilderness areas, the number of visitors each year, and the like, are useful if imperfect indicators of the policy goals of providing recreational opportunities, preserving aesthetic values, and protecting ecological systems. Other yardsticks can be used for the economic functions of natural resource policies such as ensuring the availability of sufficient rangeland, timber, minerals, water, and energy resources. Government agencies report regularly on these activities (see Council on Environmental Quality 1993), and the numbers do permit some modest assessment of important qualities of U.S. natural resources policy, its achievements, and shortcomings.

Since 1964, Congress has set aside 95 million acres of wilderness in 548 units of the National Wilderness Preservation System, and since 1968 it has designated 151 Wild and Scenic Rivers with over 10,000 protected miles (Council on Environmental Quality 1993). The Fish and Wildlife Services manages about 90 million acres in nearly 500 units of the National Wildlife Refuge System, *triple* the land area of 1970. The National Park System grew from about 25 million acres in 1960 to over

80 million acres by 1992, and nearly *doubled* the number of units in the system. By 1992, the system consisted of 362 national parks, historic sites, monuments, seashores, scenic parkways and trails, recreational areas, and memorials. The sites run from the highly popular Yosemite and Yellowstone national parks to Civil War battlefields in Virginia. The government recorded some 268 million visits to these sites in 1991, almost *double* the number in 1971. In addition, there are close to 600 million visits to national forests each year (Council on Environmental Quality 1993; World Resources Institute 1992).

If results like these suggest impressive dedication to setting aside land in a protected status and meeting the recreational needs of the American public, there are some equally troublesome statistics in the government's annual reports. The nation continues to lose ecologically critical wetlands to development. Despite the encouraging reports on growth in acreage and visits to national parks, plenty of problems remain, from congestion and crime to air pollution from nearby power plants and threats from other developments (Freemuth 1991). The Clean Air Act amendments of 1977, for example, set a national goal of preventing pollution that limits visibility in more than 150 national parks. Yet visitors are still unable to fully enjoy the spectacular vistas of the Grand Canyon and other parks due to pollution that restricts visibility and often creates unhealthy levels of ozone.

In some respects, the Endangered Species Act is typical of the halting progress and widely disparate assessments of resource policies. The federal government has listed hundreds of species as threatened or endangered, designated critical habitats, and formulated recovery plans. Yet the act has saved few species, and it is not preventing the loss of habitats and continued degradation of ecosystems (Tear et al. 1993). A good part of the explanation is a chronic and severe shortfall in budgets and staff for doing the work mandated by the act. The design of the ESA itself is also to blame, as is the tendency of the Fish and Wildlife Service to concentrate its limited resources on charismatic species that attract public support.

In 1994, for example, the American bald eagle had recovered sufficiently for the FWS to propose "downlisting" its status from endangered to threatened in most of the United States. The agency made the announcement a few days before the Fourth of July with great fanfare and wide coverage in the nation's news media. Environmental groups cited the eagle's recovery as evidence that the ESA was working and called for its renewal. Critics of the act said the rising eagle population had more to do with the nation's 1972 ban on DDT, which had depressed the eagle's reproduction rate. The National Marine Fisheries Service had announced the removal of the California gray whale from the endangered species list only two weeks previously (Cushman 1994e).

The Endangered Species Act has run into furious political opposition because it exemplifies for many the threat government regulation can pose to property rights and development projects. All the commotion is particularly striking because careful assessments of the ESA indicate that it has blocked few developments since its adoption in 1973 (Tobin 1990). A World Wildlife Fund study released in 1992 indicated that between 1987 and 1992, only 19 federal activities and projects out of 2248 submitted for formal consultation under the act resulted in cancellation due to irreconcilable conflicts between development and species protection. In the overwhelming majority of cases, federal agencies encountered no conflicts at all with the ESA. In the several hundred cases where a conflict did exist, agency officials found alternatives for moving the projects forward without unacceptable ecological damage.

Aside from such compilations of annual losses and gains in resource use, as the ESA illustrates policy conflicts continue over the core question of how to balance economic development and environmental preservation. Whether the issue is off-shore oil and gas drilling, mineral leasing on public lands, grazing rights and fees, agriculture and water use, or preservation of old growth forests, the choices often pit environmentalists against resource industries and local communities. Disputes over old growth forest ecosystems and how best to protect them have been particularly acute in recent years, with the spotted owl playing a highly visible if largely symbolic role. By some measures, only about 10 percent of the original forest remains, the rest having long been lost to timber harvests and other development.[2] That is why little choice is left at this point but to set aside large areas of forests and hence to restrict logging of them.

Unfortunately, that action has a devastating impact on small logging companies that depend primarily on use of the national forests and on the communities in which they are located. Yet other effects go unnoticed. The considerable rhetoric on job losses attributable to logging restrictions ignores distinctions between these small operations and the largest lumber companies. The latter use primarily *private* forest lands that are unaffected by laws dealing with the national forests, and those companies have enjoyed record profits as demand for housing and lumber prices soared. Moreover, the high prices for lumber have stimulated demand for alternative construction materials, including lightweight steel for framing of homes. Construction firms are finding new uses as well for unusual products to replace wood, such as baled straw that would otherwise add to the waste stream and, when burned, to air pollution. Straw houses are being built in New Mexico and other locations. States and communities also are revising building codes to facilitate the use of such materials.

These indications of program outcomes and impacts are useful, if insufficient to tell us everything we need to know about progress or the lack of it in meeting environmental policy goals. The more important point is that citizens and environmental groups should reach their own conclusions about program success and take the steps necessary to improve performance. The danger is that we may misjudge the record or the reasons for it. Like a good physician, we need to observe environmental policies carefully and be alert to a range of diagnoses and prescriptions. Generalizations about national and state trends are helpful. They are, however, no substitute for careful assessment of local situations which may differ significantly from other efforts with which they share a superficial resemblance.

COSTS, BENEFITS, AND RISKS

The improvements in air and water quality over the past 20 years, and other signs of progress in environmental and natural resources policy, are welcome news. Yet such findings do not directly address some of the major criticisms directed at those policies. One of the most important is that environmental gains come at too high a cost—in money, jobs, property rights, freedom to choose—and that alternative approaches such as market incentives or providing greater flexibility to industry and state and local governments would allow for achievement of the same environmental quality at a lower cost to society (Portney 1990b; Freeman 1990, 1994). Paul Portney has expressed the argument succinctly:

> How then do we distinguish wise from unwise policy proposals? The answer is at once very simple and very complicated. In my view, desirable regulations are those that promise to produce positive effects (improved human health, ecosystem protection, aesthetic amenities) that, when considered qualitatively yet carefully by our elected and appointed officials, more than offset the negative consequences that will result (higher prices to consumers, possible plant closures, reduced productivity). In other words, wise regulations are those that pass a kind of commonsense benefit-cost test (1994, 22–23).

According to Portney, economic research demonstrates that the nation can meet present environmental goals for perhaps as little as 50 percent of the $140 billion per year spent in 1994 to comply with federal environmental regulations. Even if the amount saved is well less than 50 percent, it would nevertheless be substantial. Such reductions can be achieved, Portney says, through an explicit, though *qualitative* and open, weighing of costs and benefits in the policy process. Some stringent environmental regulations could easily survive such a test (removal of lead from gasoline and phaseout of CFCs are two historical examples) while others might not.

These kinds of arguments for considering economic costs are difficult to dismiss. Objections certainly can be raised to relying uncritically on formal cost-benefit analysis, and environmentalists and others have done so (Tong 1986, ch. 2; Swartzman, Liroff, and Croke 1982). It is much more difficult to reject the idea that costs and benefits of environmental policies should be considered. As Portney suggests, at a minimum society has to confront some inescapable tradeoffs between environmental regulation and other forms of economic investment. Money spent on the environment is not available in the short run for other social purposes such as education or health care. Even the promise of sustainable development cannot entirely eliminate such choices however much it might point to myriad ways to better reconcile environmental protection and economic growth and to the prospect of full compatibility of economic and ecological goals in the long run.

The tradeoffs become starker as environmental management deals with marginal gains in environmental quality. That is, as we reduce pollutants to small residual amounts, the marginal dollar cost of each additional unit of improvement can rise sharply (Tietenberg 1994). The tendency of legislators to draft what economists call "absolutist" and unrealistic goals such as "zero discharge" or "lowest-achievable emissions" does not take into account such marginal costs. Nor is public support for environmental protection informed by such economic thinking. With the EPA projecting a 50-percent increase in pollution control costs between 1991 and 2000 (Council on Environmental Quality 1993), the case for reconsidering such goals is compelling.

The debate over consideration of costs and benefits extends to comparison of environmental and safety regulations. As part of its role in overseeing regulatory activities, the Office of Management and Budget has estimated the cost of regulations and calculated the cost per life saved that would result from their adoption. Using such cost effectiveness measures, some programs, such as seat belt standards in cars and benzene exposure limits for workers, are orders of magnitude cheaper than others, such as banning land disposal of hazardous waste or setting formaldehyde exposure limits for workers (Schneider 1993e; Freeman 1994).

Such a comparative exercise is potentially useful. But analysts must be able to address better than they can at present some difficult methodological problems. One of these is the necessity of considering a diversity of both acute and chronic health effects as well as the fatal outcomes that invariably are the object of attention in these studies. Another is the need to determine, and give significant weight to, the distribution of costs, benefits, and risks across the population. As the environmental justice movement has made clear, the burden of environmental pollution falls inequitably on the population, with the poor and minority groups affected more than others. Similarly, it is important to ask who pays for environmental protection efforts and who receives the benefits. Knowing only the aggregate national benefits and costs cannot possibly answer these questions, and it should not be the only basis for policy choice.

A variant of the argument for making greater use of cost-benefit or cost effectiveness analysis, as we saw earlier, is to use comparative risk assessment in which health and environmental risks are ranked to allow for the setting of policy priorities. The EPA has argued repeatedly since its initial report in 1987, *Unfinished Business,* for risk-based priority setting (U.S. Environmental Protection Agency 1990a). The General Accounting Office has as well. The core question is, if available resources are limited, how can they best be used to minimize public health risks and promote environmental quality? The premise here is that while environmental policies may reduce risks, some are far more efficient at doing so than others. Thus, depending on which programs are well funded and which regulations are aggressively enforced, governments and private parties can spend a great deal of money without a concomitant return in risk reduction.

Some environmental protection efforts could easily withstand such a comparative risk test. Examples include regulating urban smog, particulates, and lead; instituting new efforts to deal with indoor air quality; and (up to a point) limiting the buildup of greenhouse gases through reduction in use of fossil fuels. Other programs, such as cleanup of hazardous waste sites, would probably fail to measure up quite so well. Public acceptance of this approach is by no means guaranteed given conflicts between public and expert definitions of risk and a diminished public confidence in both scientific experts and government. As argued in Chapter 4, credible risk rankings of this kind also depend on improved databases and better analytic methodologies for health and ecological risk assessments (see Finkel and Golding 1994).

The effort is well underway. For example, in late 1994, a new report prepared for the California Environmental Protection Agency was praised by scientists for its careful review of evidence on dozens of environmental hazards and the risks posed by each. The two-year study by 100 scientists was especially hailed for its careful explanation of methodologies, data sources, and assumptions behind the risk assessments. The report may serve as a model for other states as the federal EPA presses them to identify and act on their most serious environmental risks (Stone 1994c).

These thoughtful and constructive efforts should be clearly distinguished from proposals debated in Congress in early 1995 that would impose far more demanding cost-benefit analyses and risk assessments on the EPA and other agencies even for fairly minor regulatory actions. A major bill backed by the House leadership, the "Job Creation and Wage Enhancement Act," would require agencies to follow an elaborate and time-consuming 23-step review process, including participation by scientific peer-review panels which could involve individuals and corporations with a direct stake in the outcome. By one count, proposals of this kind in the Republican "Contract With America" could increase the number of regulatory analyses conducted each year by 30-fold while providing little useful information to policy makers (Portney 1995). The result might well be "paralysis by analysis" as agencies struggle to meet highly prescriptive congressional demands for cost assessments (little mention is made of benefits) and other activities that offer little hope of improving environmental policy. Still, the House readily approved the bill.

Many environmentalists object on even more fundamental grounds to putting policy choices in economic terms (Sagoff 1988). In some cases, they prefer that policy debate take place on a moral rather than an economic plane. Protection of biodiversity or promotion of ecosystem health, for example, might be grounded in environmental ethics that recognize the rights of other species and future generations. Economic analysis, in contrast, tends to value the environment only in terms of its usefulness to humans. Environmentalists also question whether in any economic calculus all relevant costs and benefits will indeed be measured and weighed fairly since benefits typically are harder to estimate and society may heavily discount long-term benefits.[3] These objections are less an indictment of the principle of cost-benefit analysis than they are an expression of distrust in the way such analysis may be used by partisan advocates and of skepticism that the political process will afford an open dialogue on the critical choices.

Despite a continuing discomfort with economic analysis, environmental groups increasingly conduct their own studies as a counterpoint to government and industry estimates of costs and benefits. For example, the Wilderness Society released a report in mid-1994 contending that recreation and tourism in national forests brings in more money than timber harvests. For the area studied (five states in the Southeast), it put the value of harvested timber at $32 million a year and the economic benefits each year of recreation and tourism at $379 million. The report concluded that present policies emphasizing timber production were "out of step with economic realities" (Smothers 1994). Environmentalists have used similar studies to support arguments for protecting old growth forests in the Pacific Northwest where the region's economy is heavily dependent on tourism.

Examples of such analysis commissioned by environmental groups or government agencies abound. In an unusual three-year study focusing on the Chicago metropolitan area, the U.S. Forest Service found that planting thousands of trees would bring a net benefit of $38 million over 30 years by reducing costs of heating and cooling buildings and by absorbing air pollution. Just three strategically placed trees could save a Chicago homeowner $50 to $90 a year (Stevens 1994a).

Environmental groups in the Great Lakes area made much of recent studies on the deceptively low price of the road salt used to melt snow and ice in winter. The EPA has put the damage to vehicles, highways, bridges, and related infrastructures at $5 billion a year. A New York state agency estimated that winter road salt carried a *true* cost of from $800 to $2,000 per ton (far higher than the alternatives) when one factors in the effect on wetlands, freshwater supplies, and vegetation, among other environmental damage.

JOBS, THE ECONOMY, AND THE ENVIRONMENT

Aside from objections to the inefficiency of some environmental regulations or policies, a broader indictment has been made against environmental policy. Some scholars, politicians, and advocacy groups have implied that the nation must choose between two paths in fundamental conflict, suggested in the "economy versus the environment" debate of the early 1990s. Conservative critics of environmental policy, for example, have argued that environmental (and health and safety) regulation slows the rate of economic growth and thereby unintentionally makes the population worse off. Aaron Wildavsky (1988), for instance, asserted that by "searching for safety" and favoring governmental regulations that constrain economic investment and growth, environmentalists actually cause the population to be less safe. This is because fewer people can afford a lifestyle (e.g., diet, physical security, and access to health services) that historically has been associated with improved health and well-being. His argument hinges on the questionable assumption that regulations deal with trivial risks to public and ecosystem health and that investment in environmental protection weakens the economy and individual prosperity enough to more than offset whatever additional safety environmental policies produce (see also Douglas and Wildavsky 1982).

A special version of this argument is the "jobs versus the environment" conflict that became a significant issue in the 1992 presidential election campaign. President George Bush spoke out against "environmental extremists" and pledged to save the jobs of lumber workers in the Pacific Northwest. "It's time to put people ahead of owls," the president said, even if that meant sacrificing the nation's ancient and irreplaceable forests in Northern California, Oregon, and Washington. Candidate Bill Clinton promised that if elected, he would tackle the stalemate between lumber workers and environmentalists.

Employment and Environmental Policy

Anecdotal evidence related to the economic impact of environmental policies is easy to come by, but not very helpful. It proves little. Protection of the northern spotted owl is said to cost from 20,000 to 140,000 jobs. The American Petroleum Institute puts the blame on environmental restrictions for the loss of 400,000 jobs during the 1980s. The Motor Vehicle Manufacturers Association argued that increasing fuel economy standards would cost 300,000 jobs, an assertion repeated by Vice President Dan Quayle during the 1992 campaign. Sugar growers in Florida said protection of the Everglades would cost 15,000 jobs locally (Bezdek 1993). Mining companies claim that legislation to reform the 1872 Mining Law could cost 47,000 jobs in an industry that already suffered a 51 percent loss of its jobs between 1979 and 1992—though the Bureau of Mines estimates that only 1110 jobs would be lost (Camia 1994a).

Job losses clearly *have* occurred in specific industries, communities, and regions as a result of environmental mandates. There may be fewer losses than initial estimates often suggest, though this is of little comfort to those who lose their jobs or to the businesses forced to close their doors. Fortunately, there is much that government can do to minimize adverse impacts through job retraining programs, subsidized loan programs, and other policy interventions. Successful implementation of environmental policies requires that policy makers think more seriously about using such approaches to avoid needless conflict over otherwise broadly endorsed environmental goals.

At the national or macro level, environmental policies have only a small impact on the economy no matter what measure is used: inflation, productivity, or jobs (Portney 1990a). The reason is simple. The economy is very large (over $6 trillion in 1994) and the environmental sector is comparatively small. The one million jobs in the environmental technology industry, for example, represent eight-tenths of 1 percent of U.S. civilian employment. Thus it is scarcely surprising that economic analyses have found that environmental policies increase inflation and decrease productivity *only very slightly*. Those policies also have led to a net increase in employment (Tietenberg 1994; Peskin, Portney, and Kneese 1981).[4]

Debates over these kinds of macroeconomic impacts are largely irrelevant to the setting of environmental policy goals because the effects are relatively minor. Consistent with these conclusions, the Bureau of Labor Statistics found that during 1988, only 0.1 percent of all layoffs were attributed to environment-related causes. That is 99.9 percent of jobs lost in the United States were the result of other factors—such as freer trade, adoption of new technologies, and fundamental shifts in the economy (Bezdek 1993).

Even where employment and other economic effects are larger than these numbers indicate, economist Paul Portney offers an important observation: "counting jobs created or destroyed is simply a poor way to evaluate environmental policies" (1994, 22). A policy, and the regulations it generates, may cost jobs and still be judged desirable because it eliminates harmful pollution or protects valued resources. Conversely, policies that generate jobs may be bad for the environment and public health. We should be able to design employment policies that do not produce such negative effects.

One area that does hold much promise for job creation while simultaneously promoting sound environmental policy is development of new environmental technologies. The Organization for Economic Cooperation and Development (OECD) estimates the world market for environmental technologies at $200 to $300 billion a year, again a small market given the size of the global economy. Nevertheless, OECD forecasts sustained growth in that market over the next few decades (Gardiner 1994).

One reason for that optimistic forecast is the agreements reached at the 1992 Earth Summit. Developed nations pledged to provide a major infusion of western technology and funds to assist developing nations in designing environmentally sustainable economies. Depending on the level of financial assistance and the success of this technology transfer, the global market for environmental technologies could soar in the decades ahead.

Much the same kind of growth rate is projected for the U.S. environmental technology industry, which already consists of some 50,000 businesses. Adoption of the Clinton administration's proposed "green industrial policy" might well accelerate these trends (see Chapter 7). Such developments are consistent with a long-standing argument advanced by environmentalists—that adoption of environmentally benign technologies and encouragement of energy efficiency, source reduction, waste recycling, and similar efforts will improve employment because they tend to be more labor intensive than the polluting technologies and practices they replace (Paehlke 1994; Renner 1991).

Seeking Common Ground: Environmental Sustainability

While some of these arguments about the positive economic effects of environmental policies are long on hope and short on empirical evidence, recent studies lend support. Lax environmental standards seem to insulate inefficient and outmoded firms from the need to innovate and invest in new equipment. Such investments are likely to be essential to compete successfully in a twenty-first century global economy (Bezdek 1993). There may be some short-term economic advantages to weak environmental laws, as debates over the North American Free Trade Agreement (NAFTA) and other international trade issues have highlighted. Over the long run, however, policies that promote environmentally sound and efficient technologies and processes, not those that are environmentally degrading and wasteful, would seem to offer the best hope. This is particularly so if full cost accounting is used to fairly assess environmental damages (discussed below).

Some of the most compelling evidence of the relationship between the economy and environmental policy comes from analysis of state policy efforts. Stephen Meyer (1993) sought to test the widely accepted proposition that environmental programs adversely affect economic growth. He examined the 50 states, comparing those with strong, moderate, and weak environmental policy records in terms of employment, economic output, and productivity over a 20-year period through the early 1990s. He found that states with stronger environmental policies did not fare less well economically that those with weaker programs. In fact, he found a "consistent and systematic *positive* correlation between stronger state environmentalism and stronger state economic performance across four of the five indicators. States with stronger environmental standards tended to have higher growth in their gross state products, total employment, construction employment, and labor productivity than states that ranked lower environmentally" (Meyer 1993).

Bowman and Tompkins (1993) use different measures of state environmental policy commitment, consider different time periods, and employ a multivariate model to explain economic growth and development. They acknowledge that such analysis is difficult because of the interrelationship of the variables, and their results differ in important ways from Meyer's. Yet they too conclude that "environmentalism is not invariably associated with restricted economic growth—indeed, if anything, most of the evidence . . . favors the suggestion that it is at least *associated* with higher levels of economic growth" (8).[5]

Taken together, such findings cast considerable doubt on the validity of the "economic development versus environmental protection" argument and support the sustainable development view. President Clinton endorsed that position during his Forest Conference in Portland, Oregon in April 1993: "a healthy economy and a healthy environment are not at odds with each other—they are essential to each other" (quoted in Bowman and Tompkins 1993).

The Clinton Forest Plan for a Sustainable Economy and a Sustainable Environment, released in draft form in July 1993 and in final form in April 1994, also showed that there is much governments can do to minimize local impacts and assist in adapting to new policies. Clinton promised to spend more than $1.2 billion over a five-year period to assist the 9500 loggers likely to lose their jobs because of logging restrictions in the plan. Job retraining opportunities would be created through a new Northwest Economic Adjustment Fund. The administration argued that the jobs that would be lost immediately would be more than offset by new jobs to be produced over five years by having displaced workers repair streams and roads damaged by logging (Camia 1993a; Egan 1994a).

Ironically, even before the promised federal funds began to reach Oregon, the state was reporting the lowest unemployment rate in a generation and rising wages and property values. Between 1989 and 1994, the state lost 15,000 jobs in the forestry product industry. However, it *gained* nearly 20,000 jobs in high technology, such as computer manufacturing. By 1995, technology is expected to surpass timber as the leading source of jobs in the state. Even the most timber-dependent communities in the state saw a net increase in jobs and rising property values (Egan 1994b).

In some cases, short-term conflict between environmental policies and the economy cannot easily be reconciled, and policy implementation may have to be altered or alternatives sought to maintain public support. A good example is clean air enforcement in Southern California in the early 1990s when the area was suffering from a severe recession, worsened by deep defense spending cutbacks. In 1993, the South Coast Air Quality Management District (SCAQMD) was forced to slow the pace of implementation when local industry and government complained bitterly about the costs and burdens it was imposing on them. SCAQMD saw its budget cut and its staff reduced by some 20 percent between 1990 and 1993. Public resentment over lost jobs and mandatory carpooling policies for larger employers caused at least some of the backlash. The agency's director was moved to remark that "the economy is changing priorities" (Reinhold 1993a).

REFORMING ENVIRONMENTAL REGULATION

A persistent criticism of environmental policy is that it relies too heavily on centralized, technically driven, "command-and-control" regulation in which government sets environmental quality goals, methods to achieve them, and deadlines, with penalties for noncompliance. Aside from the issue of costly inefficiencies addressed above, such regulation is often said to be poorly conceived, cumbersome, time-consuming, arbitrary, and vulnerable to political interference. Some add to the litany of complaints that Congress cannot seem to resist the temptation to add pork-barrel ingredients to the regulatory recipe—for example, funneling large sums of money to municipal sewage treatment plants that are overbuilt for the populations they serve. Hence in recent years, economists and policy analysis have suggested a range of alternatives that may either substitute for regulation or supplement it (Freeman 1994; Stavins 1991).

The Case for Policy Alternatives

There is little real argument today over the wisdom of using such alternatives to regulation as market-based incentives, provision of information, public education, negotiation, voluntarism, and public-private partnerships. They are among the policy approaches that governments at all levels consider, and they have enjoyed wide use in the 1990s. Many environmental problems simply cannot be addressed effectively with traditional regulatory tools alone, but incentives and educational efforts may be useful supplements. One example is nonpoint water pollution control that must deal with thousands of dispersed sources such as small farms. Another is indoor air quality, which affects people in millions of individual homes and commercial buildings. A third is the use of product labels and other forms of communication to inform consumers about the energy efficiency of appliances, the content of recycled materials, or other attributes related to environmental quality or public health. "Green labeling" has been widely used in Europe and will soon become more visible in the United States.

There are, however, questions that we need to ask about how effective these policy alternatives have been or are likely to be. What are the advantages and disadvantages of their use? What about the political and institutional implications of any wholesale shift away from regulation and toward such alternatives? Some empirical studies suggest, for example, that provision of information through product labels is an ineffective means of changing consumer behavior. People often do not read the labels or understand or use the information they contain. Such findings have stimulated a search for better labels that may enhance public understanding and use (Hadden 1986).

The Promise of Market Incentives   The basic logic of market alternatives is clear enough. If, for example, motorists drive too much and waste fuel, causing urban congestion and pollution, and contributing to the buildup of carbon dioxide in the atmosphere, how could such behavior be discouraged? Various regulatory schemes might work, from setting high fuel economy standards for automobiles to restricting use of private vehicles through carpooling and other means. Yet they would involve a level of bureaucratic intrusion that may be socially unacceptable. They also have not worked very well. Although the average automobile is far more efficient today than 20 years ago, public preferences recently have favored decidedly inefficient vans, light trucks, and higher performance vehicles; 40 percent of vehicles sold today are light-duty trucks and vans.

A market alternative would be to raise gasoline taxes sufficiently (those in the United States are among the lowest in the world) to achieve the same goals. Higher prices for gasoline will create incentives for motorists to drive less, seek more fuel-efficient vehicles, and rely more on mass transit. Changing consumer preferences will build a market for efficient vehicles to which automobile companies can respond. In theory, a market incentive such as this one can reduce the bureaucratic burden associated with regulation. Regrettably, public resistance to higher fuel costs and timidity on the part of public officials makes imposition of a steeper gas tax (or a broader carbon tax) politically infeasible at this time.

Both the EPA and the states have experimented over the past two decades with market incentives and other alternatives to regulation. These innovations are encouraging, if still controversial and limited in actual use. Some of the earliest applications of the approach were in air pollution control where offsets and "bubbles" were created to introduce greater flexibility and promote economically efficient solutions. Offsets allow new sources to locate in nonattainment areas if other sources in the area can reduce their emissions. The bubble concept treats a facility as if its various emissions were under a single umbrella or bubble. Firms may establish a new pollution source by reducing emissions elsewhere in their facility. Congress went further in the 1990 Clean Air Act when it authorized emissions trading to reduce precursors of acid rain, sulfur dioxide and nitrogen oxides.

The EPA began to make use of market incentives like this in 1976 (Council on Environmental Quality 1993, 56–57). In addition to the offset and bubble policies for air pollution control, programs in which incentives were used included wetland mitigation banking, the phaseout of lead in gasoline, an air toxics offsets program, scrappage of old cars ("cash for clunkers"), privatization of wastewater systems, and safer pesticides incentives.

Innovation in State and Local Governments   DeWitt John (1994) examined similar innovations at the state level. He shows how through the use of nonregulatory tools and cooperative approaches, state governments reduced the use of agricultural chemicals in Iowa, helped to devise a plan for restoring the Florida Everglades, and used demand-side management to conserve electricity in Colorado. In a related effort, in 1993, as the South Coast Air Quality Management District sought to minimize adverse economic impacts of its clean air program in Southern California, it adopted market incentives as a less costly way to achieve clean air goals. A local market in pollution credits (the Regional Clean Air Incentives Market or RECLAIM) replaced dozens of rules. Overall industrial emissions must be reduced each year and achieve a total cut in nitrogen oxide emissions of 75 percent by 2003, and 60 percent in sulfur dioxide emissions from current levels. Nearly 400 large companies have been given flexibility to find the cheapest way to achieve those reductions; they may trade pollution rights within a declining cap on total emissions (Reinhold 1993a).

Other examples suggest the diverse uses of market incentives and similar alternatives at the state and local level. For reasons discussed in Chapter 4, local governments are hard-pressed to treat growing quantities of wastewater and to meet rising quality standards for wastewater that flows into streams, rivers, and bays. Cities across the nation have concluded they are better off extending rebates to residents willing to install new toilets than treating the extra wastewater that would otherwise be generated. New models being marketed under EPA rules that took effect in January 1994 use only 1.6 gallons of water per flush in comparison to over three gallons for the older ones.

Another example of such innovation is in energy conservation, as discussed in Chapter 5. Utilities in the early 1990s offered discounts of 80 percent or more to customers buying energy-efficient light bulbs. It was cheaper for the companies virtually to give away the new bulbs than to build the extra generating capacity they would need to meet consumer demand. The regulatory and financial hurdles of designing, getting approval for, and building additional plants made the light bulb offers attractive in comparison. Much the same is true of rebates offered by utilities to buyers of energy-efficient appliances.

Even with market incentive systems, much of the activity now required under regulatory policies is still needed. Environmental quality criteria and standards must be set, and monitoring and compliance actions continued. The most important difference lies in giving state and local governments and industry greater flexibility in the means they use to achieve the goals. In some cases, market incentives may not work as well as regulation, and they may not achieve environmental goals as quickly. Much depends on how they are designed and implemented. Where such incentives fail to do the job, governments always have the option of returning to mandatory standards and regulation.

Assessing New Policy Approaches

Some skepticism is warranted in appraising these new policy instruments, particularly when experience with them is limited. Just as environmentalists and policy makers favored regulatory policies in the 1970s and 1980s, policy debate in the 1990s is dominated by the language of business schools. Policy analysts urge that programs be more customer driven, decentralized, and competitive, and that citizens, communities, and bureaucrats be empowered to take action in an entrepreneurial spirit. Mission-oriented problem solving and flexibility are to replace fixed rules and procedures, and design and market incentives are favored over planning and regulation. Privatization and contracting out are touted as more effective than relying on government employees. Many of these proposals are problematic while others may be reasonable supplements or alternatives to current practice. Few, however, have been studied carefully.

Policy analysis and program evaluation can help separate the promising approaches from the unworkable. For which environmental programs does decentralization make sense, and for which does centralized, uniform, national policy work best? To what extent can reliance on private lawsuits reduce the need for government regulation, for example, in oil-spill prevention?

Tort Law and Natural Resource Damage Assessments   One illustration of the potential of using tort law comes from the 1989 Exxon *Valdez* oil-spill cleanup and settlement. As of fall 1994, Exxon had spent $2.1 billion cleaning up Prince William Sound in Alaska and $1.3 billion in civil and criminal penalties and settlements of claims filed by some 11,000 residents and businesses in the area. In August 1994, a federal jury ordered Exxon to pay $286.8 million in compensatory damages to over

10,000 Alaskan fishermen who had filed a class action lawsuit. The biggest judgment by far was another jury decision in September 1994 ordering the oil company to pay $5 billion in punitive damages to about 34,000 fishermen and other Alaskans. The huge award, which Exxon vowed to appeal, was by far the largest civil judgment ever in a pollution case. It nearly equaled Exxon's net earnings of $5.3 billion in 1993 (Schneider 1994d).

As a result of recent court decisions, the Interior Department and other agencies are making more use of natural resource damage assessments. Some of those will include measurement of the loss of so-called "passive uses" of the environment by citizens not directly affected, a move environmentalists applaud. Economists use so-called contingent valuation methods (indirect measures of valuation) that rely on public opinion surveys to estimate those values, for which there is no market equivalent. If widely used to assess damages under federal environmental laws, such methods could dramatically alter the willingness of corporations to engage in high-risk endeavors for which they could be held financially liable (Passell 1993).

Contracting Out and Privatization   The use of market incentives and tort law is still limited. These approaches may turn out to work far better in some instances than in others. It would be difficult to generalize about their promise. Instead, for these and other alternatives to regulation, policy analysts will need to determine which problems are most amenable to resolution through such approaches and which are poor candidates. A few examples of contracting out and privatization suggest what needs to be examined.

Consider the tendency of some government agencies to rely heavily on outside contractors, a form of privatization. How well have the EPA, the Department of Energy, the Department of Defense, and other agencies managed their contractors? What are the savings, or costs, associated with such programs? What other impacts are important?

Some of the evidence to date gives cause for concern. Donald Kettl, for example, reports that the EPA was so dependent on contractors for its Superfund program that it turned to them for help in responding to congressional inquiries, analyzing legislation, and drafting regulations and standards. Contractors "drafted memos, international agreements, and congressional testimony for top EPA officials. . . . They even wrote the Superfund programs's annual report to Congress" (Kettl 1993, 112). Cases such as this have led the General Accounting Office to be sharply critical of the EPA's lax supervision of contractors, which made the agency vulnerable to contractor waste, fraud, and abuse (U.S. General Accounting Office 1992).

Much the same could be said for many DOE programs. For example, about 85 percent of the 2000 people working on DOE's Yucca Mountain, Nevada, nuclear waste site investigation are contract, not departmental, employees. The General Accounting Office has also criticized the DOE for inadequate supervision of contractors that costs the federal government millions of dollars a year (U.S. General Accounting Office 1992). Under these conditions, we have to ask not only whether any financial savings are likely to accrue to taxpayers but whether political accountability for these programs has been sacrificed. In July 1994, Secretary of Energy Hazel

O'Leary announced a new process of competitive bidding on DOE contracts that she hoped would address such concerns. Emphasis was to be placed on smaller, better-defined contracts and improved cost controls (Cushman 1994f).

How well have other kinds of privatization efforts worked out? In one celebrated case, Los Angeles offered sizable performance bonuses for early completion of free-way repair following the 1994 earthquake. The private construction companies finished the work faster than anyone believed possible. Representing a different kind of private action that is gaining favor in environmental circles, since 1951, the Nature Conservancy has purchased ecologically sensitive habitats to help preserve species. It now is responsible for over five million acres worldwide. The conservancy also works with developers, industry, farmers, and local governments to encourage new approaches to resource management. Other environmental groups have pursued similar strategies.

Despite its poor reputation during the Reagan administration's efforts at environmental deregulation (Vig and Kraft 1984; Portney 1984), the case for privatization of public lands has not entirely vanished. A small group of economists advocating "new resource economics" have persistently argued that private ownership of public lands will increase efficiency in management and, through the market's invisible hand, yield an optimal mix of goods and services for society. They believe that even noncommodity values such as wilderness and recreation, will be better served in private markets (Stroup and Baden 1983; Anderson and Leal 1991). Other environmental economists do not share their faith in the ability of free markets to protect ecosystems and to advance other public values such as equitable access and preservation of resources for future generations.

One of the unanticipated consequences of deregulation in the electric utility industry may serve as a warning to treat privatization and deregulation proposals carefully. In September 1994, Ivan Selin, the chairman of the Nuclear Regulatory Commission, suggested that new market pressures were creating an "incentive to cut corners" at nuclear reactors in order to compete in the emerging free market economic environment. Whether such competition will actually lead nuclear plant executives to compromise safety measures is a subject of debate in the industry. Skeptics suggest that such decisions are unlikely given the exceptionally high cost of nuclear plants. They believe most utilities would find it cheaper to invest heavily in maintenance to keep the plants running. Nonetheless, it is instructive that the NRC has indicated that its staff will be looking carefully at this "new area" of concern, particularly at plants with poor operating records (Wald 1994b).

Each case for privatization or greater reliance on competitive markets, of course, must be assessed on its own terms. Markets and incentives may work well in some instances and not in others. There is no persuasive case for across-the-board reductions in government budgets and staff (as emphasized in the 1980s) and a switch to market mechanisms on the grounds that markets are inherently superior for allocation of public goods. One must remember also that markets must be structured in some fashion and policed, and consumers must have access to the necessary information to make rational choices. These requirements imply a continued need for government bureaucracies and the courts.

NEW DIRECTIONS IN ENERGY AND NATURAL RESOURCE POLICY

As several of these examples and the discussion in Chapter 5 illustrate, many of the same ideas prominent in discussions of environmental protection policy are proposed for reform of natural resource and energy policies. Critics of all stripes have long found fault with policies they consider outmoded, costly to the government, and environmentally destructive. These distributive or subsidy policies include extensive federal support of commercial nuclear energy from its inception in the 1950s, below-cost sales of timber by the U.S. Forest Service, and the Army Corps of Engineers' long-standing preference for building dams and levees as the chief way to control flooding. Other prominent examples are minimal grazing and mining fees for use of federal land in the West that have led to degradation of land and water supplies and lost revenue to government. Most of these policies had strong support in the U.S. Congress and powerful constituencies prepared to defend them, making their elimination politically difficult in any administration. Yet the political climate of the 1990s suggests that the policies will continue to be critically examined and possibly eliminated (Davis 1991; Egan 1993b).

One effort to do so occurred in early 1995. A broad coalition of environmental groups, advocates of free market economics, and conservative taxpayer organizations identified nearly three dozen federal programs that harm the environment and may cost the nation as much as $33 billion over a 10 to 15-year period. They named public land subsidies, disaster insurance, subsidized irrigation for farmers, and foreign aid projects, among others. The Green Scissors coalition urged the new congressional majority to eliminate the programs as part of its campaign to cut government waste and inefficiency (Lancelot and De Gennaro 1995).

The Logic of User Fees

Among the most important natural resources policy reforms are imposition of higher fees for those using public lands. Additional fees bring more money to the federal treasury, though that is not the main reason for instituting the changes. The purpose is to eliminate inequitable subsidies to user groups, especially for programs that degrade the environment and deplete or waste natural resources, and to increase market efficiencies.[6] The federal government spends about $17 billion a year to manage resources that produce less than $7 billion a year in revenues, and it also must cover the cost of repairing environmental damage. Those costs may be quite high. For example, some $11 billion will be needed to clean up abandoned mines that threaten groundwater, primarily in the West.

In perhaps the most egregious case under the 1872 Mining Law, Secretary of the Interior Bruce Babbitt was forced by a federal court ruling in May 1994 to sign a contract with a Canadian-based mining company, American Barrick Resources, for a mere $9565. That contract gave the company the right to mine an estimated $10 billion worth of gold on public land without paying any royalties on the gold (Cushman 1994c). Publicity over the case strengthened the hand of those urging reform of the 120-year-old law.

The General Accounting Office has estimated that $1.7 billion in hardrock minerals is taken each year from federal land. Under proposed legislation in the House, miners would pay an 8-percent royalty on extracted minerals. The Senate bill considered in 1994 was much less demanding on the miners and reflected aggressive lobbying by the American Mining Congress. Although Congress failed to approve new legislation in 1994, the issue is likely to be addressed in the new session of Congress.

There is no shortage of examples of comparable inequities in natural resource policies. The Council on Environmental Quality (1993) reports that about 40 percent of Bureau of Land Management rangeland is in poor or fair condition, and only about 5 percent in excellent condition. Low federal fees charged to rangeland users are widely cited for contributing to overgrazing and environmental degradation. From this perspective, the fees government charges should roughly approximate the actual worth of the resources that are used or the environmental damage that is caused. Market forces would then discourage environmentally destructive behavior rather than encouraging it as present policies do.

Social Cost Accounting

An even more important step to take advantage of market forces would be the adoption of economic accounting measures that reflect the depletion or degradation of natural resources and thus more accurately depict the country's standard of living and its sustainability. The Council on Environmental Quality (1993) reports that agriculture, forestry, fisheries, and mining combined contributed about $128 billion to the nation's $5.7 trillion economy in 1991. Yet such accounts have not reflected the obvious loss that occurs if nonrenewable resources are exhausted or damaged. The idea of such environmental or social cost accounting has circulated among economists and environmentalists for years. It is finally being taken seriously by public officials in the United States and globally (Starke 1990; Lutz 1993). In 1993, for example, President Clinton directed the Bureau of Economic Analysis in the Commerce Department to begin work on recalculating the U.S. Gross Domestic Product along greener lines. During Earth Week in 1994, the bureau announced its intent to release the first limited data using such measures, which it designated "integrated economic and environmental satellite accounts." It also promised more extensive environmental accounting over the next several years.[7]

Similarly, through its Design for the Environment Program, the EPA is working with industry, academia, and the accounting profession to encourage a fuller and more accurate consideration of environmental costs. The United Nation's Statistical Commission is developing international guidelines for similar measures. In a related move, the National Oceanic and Atmospheric Administration (NOAA) and the Department of the Interior are formulating regulations to measure natural resource damages that result from release of oil and other hazardous substances to the environment. Such assessments facilitate collection of damages from responsible parties.

Government Purchasing Power

Another alternative to regulation is using the buying power of government to change existing markets. The federal government alone buys an estimated $200 billion a year in goods and services. If added to purchases by state and local governments, the public sector accounts for 18 percent of the nation's total economic output. Some simple changes can produce remarkable effects on the economy and on emerging "green markets" (Jacobson 1993).

A good example of using purchasing power concerns computer equipment. Computers—including copiers, fax machines, and printers—have been the fastest growing energy users in commercial buildings (and colleges and universities) in recent years. Effective in late 1994, all federal agencies must buy energy-efficient computers when they replace equipment. The EPA estimates that purchase of the "Energy Star" rated computers and printers will save the government some $40 million each year in electricity costs (National Performance Review 1993b; Regan 1993a). The federal government accounts for only 3 to 5 percent of national computer sales. However, by guaranteeing a market for energy-efficient machines, the policy allowed manufacturers to shift to new technology. That change involved the minor redesign of monitor, printer, and computer circuits. Reduction in electricity use nationwide should be substantial. EPA estimates put the savings from efficient printers alone at up to $450 million per year.

The federal government also has mandated that all agencies use paper made from at least 20 percent recycled fiber (increasing to 30 percent by 1998). The Government Printing Office already uses recycled paper for most congressional documents. It prints the *Federal Register* as well as the *Congressional Record* on 100 percent postconsumer recycled paper. Although it buys less than 2 percent of the nation's printing and writing paper, the federal government's policy on use of recycled paper will serve as a model for state and local governments as well, especially since environmentalists vowed to lobby for wider use of the federal standard.

Other areas targeted by the Clinton administration for similar actions included the purchase of alternative fuel vehicles and CFC-free air conditioning systems, and use of vegetable-based inks in printing. In late 1994, Congress approved legislation requiring the federal government to use soybeans and other vegetable oils and materials derived from renewable resources when technologically feasible. Designed to help farmers as well as to protect the environment, the measure set varying standards for ink used in government printing, depending on the type of printing process. Ink used on newsprint, for example, would have to contain at least 40 percent vegetable oil.

Ecosystem Management

A more fundamental way to reform natural resource policies as well as pollution control efforts is through the use of ecosystem management. Historically, governments developed policies to deal with discrete elements of the environment such as forests, water, soil, and species, often through different federal agencies and without coordination among them. The EPA for two decades dealt almost exclusively with public health, and paid little attention to maintenance of ecosystem health. The agency rarely coordinated efforts across the separate media of land, water, and air. The conventional approaches have contributed to ineffective and inefficient environmental policy.[8] They also have resulted in public policies that concentrate limited resources on small public health risks that bring few benefits at high cost while larger risks are ignored. The backlash against environmental policy in the 1990s—from industry, state and local governments, and the property rights and Wise Use movements—derives from these policy weaknesses as much as anything else.

Within the last few years, the more comprehensive ecosystem management approach has gradually established a foothold. It reflects advances in ecological research that are beginning to build an understanding of ecosystem functioning. To some extent it also incorporates ideas of ecological rationality advanced by social scientists (Dryzek 1987; Bartlett 1986; Costanza, Norton, and Haskell 1992). Rather than focus on individual rivers and localized cleanup efforts, for example, emphasis shifts to entire watersheds and the diverse sources of environmental degradation that must be assessed and managed. Instead of trying to save individual endangered species, the new goal is preservation or restoration of ecosystems themselves.

New approaches to ecological risk assessment, although still in their infancy, promise to provide the necessary knowledge on which to base a broader, more holistic, and better focused attack on environmental problems. Such assessment can create a systematic process for clarifying risks and setting priorities (U.S. Environmental Protection Agency 1992a). That is one of the purposes of Interior's Biological Service—to inventory the nation's biological resources and strengthen knowledge of threatened habitats and species to facilitate better management of them (Stone 1993).

Carol Browner has begun moving the EPA toward use of ecosystem management and related comprehensive approaches. Similar efforts are underway in the natural resource agencies. Browner recently announced, for example, a major initiative to explore—on an industry-by-industry basis—coordinated rule making, multimedia compliance and enforcement, and other approaches that offer "cleaner, cheaper" environmental results (Gardiner 1994).

Similarly, in 1992, the Interior Department's Bureau of Land Management and Forest Service announced forestry initiatives based on ecosystem management that would reduce clearcutting as a standard practice. Institutional and political obstacles to using ecosystem management and other comprehensive approaches to environmental decision making remain, as opposition to the biological inventory in Congress reveals. Nevertheless, the trend is encouraging, and it suggests an important step forward in the evolution of natural resources and environmental policy. The Clinton administration's 1995 budget request included $700 million for ecosystem management initiatives.

Comparable efforts are apparent in the seemingly more mundane arena of the Army Corps of Engineers. A major catalyst here was the devastating Mississippi River floods of 1993, for which a major contributing cause was the extensive system of dams and levees along the river that the corps had constructed. The Clinton White House charged the Secretary of Agriculture, Mike Espy, with reviewing flood control programs and recommending actions to prevent similar damage in the future.

After wide consultation with governors, mayors, community residents, and scientific experts, a study committee appointed by Espy recommended in mid-1994 that greater efforts be made to promote evacuation of flood plains and relocation of businesses and farms to higher ground, assisted with federal financial incentives. In the long run, the nation would save money by avoiding costly emergency relief efforts. As a key component, the report recommended restoration of natural hydrologic cycles that had been severely disrupted by eliminating wetlands in the river's basin.

This report relied heavily on the idea of "watershed management," which considers the impact on a river of the wider basin of which it is merely a component. In this form of ecosystem management, the prevention or mitigation of flooding depends on preserving and restoring wetlands in addition to whatever maintenance of engineered flood control mechanisms is needed. Acting somewhat like a sponge, wetlands absorb excessive rain and release it slowly, thus limiting the risk of flooding. Wetlands perform other vital ecological functions as well (Cushman 1994d).

CONCLUSION

Over the past two decades, environmental policies have become highly complex and their reach has extended to every segment of the economy and every corner of the nation. Their costs and impacts on society are much debated. They are also at the center of many political controversies over the role of government in protecting public health and managing the nation's natural resources. Evaluating their successes and failures helps to inform these debates even if such studies cannot answer all questions or eliminate conflicts over the direction of environmental policy.

One of the best ways to ensure that new scientific knowledge and policy and program evaluations are brought to bear on policy choices is to design such studies into the policies themselves. The Montreal Protocol that governs international action to phase out ozone-depleting chemicals like CFCs offers one model for doing so. It integrates continuous assessment of changing environmental conditions with new policy actions. Congress occasionally has provided mechanisms of this kind for pollution control policies. In 1977 it created the National Commission on Air Quality and charged it with overseeing and evaluating the EPA's performance in implementing the Clean Air Act. The 1972 Clean Water Act created a National Commission on Water Quality to study technical, economic, social, and environmental questions related to achieving certain goals established by the policy. Its report guided Congress in revision of the act in 1977 (Freeman 1990).

Whether through these kinds of devices or other mechanisms, environmental policy in the 1990s would benefit from regular and systematic assessment of progress. Such evaluation must look at measurable program outcomes. For reasons discussed in this chapter, however, that alone would be insufficient. Equally important is assessment of long-term impacts on institutions and decision-making processes within and outside of government.

Another imperative is rethinking the logic of policy design (Schneider and Ingram 1990). For policy to work well, it must be based on an understanding not only of environmental conditions but of the people and institutions who implement the policy and those who have to comply with the regulations and other directives it produces. We can and should do a better job in the future. Citizens can help by informing themselves about how well policies are working and participating in decision making at all levels of government that affects the reform and redirection of environmental policy.

ENDNOTES

1. The general line of argument can be seen in a special issue of *Regulation* (vol. 17, no. 1, 1994) dealing with environmental regulation. *Regulation* is a quarterly publication of the Cato Institute that offers critiques of health, safety, and environmental regulations and suggestions for their reform. An even stronger and more ideological critique emerged among conservatives in the late 1980s. It was reflected in a widely publicized speech at Harvard University in May 1990 by former Bush OMB director Richard Darman. He denounced the "radical, anti-growth green perspective" for its "absolutist approach to environmental values." Darman asserted that the label "environmentalist" was "a green mask" that could conceal certain ideologies inimical to U.S. interests (Darman 1990). The same tone can be seen in the conservative press, where some writers appeared to find in environmentalism the kind of threats to traditional American values they formerly detected only in radical left movements (Horowitz 1990).

2. The Forest Service asserts that more than six million acres of old growth forest remain in Oregon and Washington. The Wilderness Society, using a different method, says the real number is only 2.3 million acres (World Wildlife Fund 1994). The 10 percent figure comes from the Wilderness Society study.

3. The concern is legitimate. Yet economic methods for estimating environmental and resource values are fairly sophisticated and widely used. For a discussion of how benefits and other environmental values can be estimated and used in decision making, see Freeman (1993). For a contrasting perspective, see Sagoff (1993).

4. Estimates of economic impacts show a wide variance, depending on the methods and models used, and on the economic indicators selected. Over the past decade, such studies indicate that the impact on inflation and productivity is probably between 0.1 and 0.4 of a percentage point. In one early accounting, the CEQ noted in its 1979 *Environmental Quality* report that the annual accruing benefits of all federal environmental programs were substantially greater than the total cost of compliance. It also found the impact of these programs on the economy to be marginal.

5. Consistent with these findings from state-level analysis, Bezdek (1993) reports some evidence that nations with the most stringent environmental policies have the highest rates of economic growth and job creation.

6. The GAO reported in 1994 that studies by agricultural economists suggest that higher water prices being instituted in western water projects will increase irrigation efficiency and conservation, and thereby reduce environmental degradation attributable to irrigation as well as free up water currently used for irrigation for other purposes (U.S. General Accounting Office 1994b).

7. By some accounts, these initiatives date back to proposals during the Ford administration. The Commerce Department formally began the new calculations in 1992, the last year of the Bush administration.

8. The Conservation Foundation, now part of the World Wildlife Fund, for years sponsored research and issued position papers on cross-media pollution control as part of its Options for a New Environmental Policy Project. One element of that work was the development of a model environmental policy for the nation that incorporated these integrative principles. See Rabe (1986) and National Commission on the Environment (1993).

CHAPTER SEVEN

Environmental Policy and Politics for the Twenty-First Century

An assessment of environmental policy and politics requires that we look ahead to determine public needs for the twenty-first century. Several visions and forecasts provide some light to guide the way, none without bias and all with limitations that come with the territory. Some analysts, such as Julian Simon and the late Herman Kahn (1984), have projected a future of economic prosperity and technological advancement, with few environmental problems that could not be solved with human ingenuity and determination. Environmentalists maintain that the forecasts are inaccurate and the arguments unconvincing. They cite other projections and analyses that find ecological imperatives will force humanity to adjust to the physical realities of the biosphere (Meadows, Meadows, and Randers 1992; Ophuls and Boyan 1992; Council on Environmental Quality and Department of State 1980). Such studies urge profound, sometimes wrenching, changes in human behavior and institutions to avert potential disasters. Not all are convinced. The skeptics believe such work discounts too heavily the potential for scientific and technological invention and the human capacity for adaptation.[1]

These differences in outlook underscore a simple truth. The future is shaped by more than specific demographic, technological, and environmental changes. It also reflects human perspectives about the natural world and the personal and political choices we make. Policy makers are beginning to realize that. For example, UN Secretary-General Boutros Boutros-Ghali has formed an advisory committee to integrate global changes into a vision for the twenty-first century. The committee will pay special attention to "paradigm shifts" in our basic assumptions about human civilization and the way we perceive and think about changing environmental and other conditions.[2]

We may disagree about the likely character of society 50 to 100 years in the future, but uncertainty about technology, social trends, and the environment need not prevent us from acting now. Climatologist Stephen Schneider (1990) has argued that in response to environmental threats such as global warming, society chooses among three quite different strategies: (1) reliance on technological fixes or corrective measures with little change in human behavior and institutions; (2) adaptation to changing environmental conditions without attempting to counteract them; and (3) preventing the adverse changes from occurring by altering the practices that cause them. Environmentalists prefer the preventive measures, which they believe are essential to avoid severe or irreversible consequences. Such policy measures are seen as a form of insurance to prevent or mitigate possible harm.

Debate over environmental policy reflects contrasting judgments about the wisdom of such different policy strategies and the social, economic, and political values they represent. Environmental policies have been based on each of the three strategies, in sometimes awkward combinations, with an increasing emphasis on prevention of environmental degradation and threats to public health. Even as conflict continues on many policy issues, considerable agreement is evident on the broader environmental agenda for the twenty-first century.

ENVIRONMENTAL GOALS AND POLICY CHOICES

Any review of global environmental conditions and trends at the end of the twentieth century suggests that the United States and other nations are at a historic crossroads. The 1992 Earth Summit explored at length the nature of the world's environmental problems. The Agenda 21 that meeting produced set out an ambitious blueprint for sustainable development and for the institutional and policy changes it requires. Instituting those changes will preoccupy nations of the world over the next several decades and possibly for the entire twenty-first century. How the United States and other countries respond to those challenges, how swiftly we devise suitable environmental policies and other reforms, will significantly affect the quality of life we enjoy in the future. The nation is slowly coming to terms with that reality, but it will not abandon old ideas, policies, and practices easily.

As the previous chapters have shown, U.S. environmental policies have produced significant gains over the past 25 years. They promise to bring further improvements in environmental quality and human health and well-being. It is equally evident that those policies and their implementation have exhibited profuse, though not fatal, weaknesses. As Chapter 6 argues, policy makers at all levels of government need to design more efficient and effective policies and programs. Where necessary, they need to consider alternatives to conventional regulation such as market incentives, public education, and public-private partnerships. They also should do a better job of setting priorities to make the most of scarce fiscal resources. In addition, they need to reduce or eliminate the adverse impacts of environmental policies and to help reconcile environmental protection activities with other social and economic goals.

Those are the comparatively easy tasks. The tougher job is building institutional capacities to stimulate and maintain sustainable development and fostering the changes in society's values that are as essential to environmental sustainability as public policies. Nationally and internationally, we need to recognize, assess, and respond effectively to dynamic and uncertain environmental problems even when political consensus is elusive and social and economic conflicts divide us. Knowledge of environmental problems and the effects of public policies often is insufficient. Yet just as important is the failure to use what we do know. We disagree about policy goals, priorities, and timetables for action, and our political processes facilitate a thorough airing of those conflicts. Unfortunately, they have proved less able to build consensus around viable environmental policies.

Of particular importance are those institutional processes that relate to the use of scientific knowledge and the creation of a public dialogue on environmental risks that can help to improve policy decisions. Scientific knowledge grows more rapidly than our ability to digest and act on it, and we have yet to discover the best way to assess that knowledge and bring it to bear on the policy process. Any solution depends on involving the public in a dialogue on the issues. There are no alternatives in a democratic political system. That does not make it any easier, however, to find ways in which to engage the public that can enhance rather than diminish the quality of that discourse and the effectiveness of the policies the nation chooses.

These policy and institutional needs, and the imperative of public education and participation, define in broad outline the environmental agenda for the 1990s and twenty-first century. Those eager to play a role in shaping these developments and in devising solutions to environmental problems will find a rich diversity of analyses and prescriptions to assist them.

An Evolving Policy Agenda for the 1990s

Environmental policies consist of governmental responses to a moving target. As problems and society's definition of them change, policies require reconsideration and adjustment. In one sense, then, the environmental policy agenda is in a constant state of flux. Responses to these changing needs are sometimes simple and fast. More often they are complex and slow because adjustments involve dealing with scientific and economic uncertainties and resolving political conflicts. In both cases, policy makers increasingly have favored an eclectic mix of environmental policies, employing regulatory as well as less conventional means. The trend is encouraging.

Reforming Environmental Policies and Institutions

As Chapters 4 through 6 illustrated, the last few years have witnessed a barrage of criticism of environmental policy from all sides—industry, state and local governments, property rights and Wise Use advocates, and environmentalists themselves. Congress and state and local governments have been rethinking policy goals and means, particularly when renewal of the major acts presents them with the opportunity to do so. In 1994 alone, Congress was busy reassessing the Clean Water Act, the Safe Drinking Water Act, the Resource Conservation and Recovery Act, Superfund, FIFRA, and the Endangered Species Act. Members also considered reform of the 1872 Mining Law and Clinton administration proposals for changing use of public lands and other natural resources. The 103rd Congress was unable to agree on new directions for these acts and the process began anew in the 104th Congress, which was far more critical of environmental policy.

Most of the proposed policy changes considered in 1994 represented incremental adjustments to environmental statutes that survived an elaborate process of policy legitimation—involving extensive hearings and negotiation among members of Congress as well as environmentalists, the business community, and state and local government officials. This was not the case, however, for many proposals acted on in the 104th Congress, including the regulatory reform provisions of the Republican "Contract With America" as noted in Chapter 6. Given the political climate on Capitol Hill in 1995, members seem likely to approve most of those proposals, though possibly with amendments that will soften their adverse impacts on environmental policies. Especially worrisome was the price likely to be paid for rushing to enact radical policy changes, most notably in the House under the leadership of Speaker Newt Gingrich. The tight legislative schedule provided few opportunities to assess the consequences of policy decisions for public health and the environment.

The remarkable efforts in the 104th Congress to weaken environmental policy serve as an important reminder of the degree to which politics and ideological judgments may shape policy choices. The congressional eagerness to approve the "Contract" occurred despite public opinion polls that continued to demonstrate strong public support for environmental protection. Members of Congress backing these proposals also ignored the broad consensus that had formed over needed reforms in environmental policy. Improving the efficiency of environmental regulations, employing comparative risk assessment as one basis for setting policy priorities, making greater use of market-based incentives, providing additional flexibility to promote compliance on the part of industry and state and local governments, and giving new weight to pollution prevention, are now widely, if not unanimously, endorsed (National Commission on the Environment 1993; Portney 1995). As congressional action in early 1995 vividly indicated, however, much conflict remains over the precise way in which such reforms should be formulated. That continuing conflict greatly constrains the nation's capacity to make intelligent choices about environmental policy.

Similar barriers affect the degree to which other promising proposals that have been much debated and studied over the past decade might receive the requisite political endorsement and thus help to improve environmental policy. These include the use of comprehensive and integrated approaches to environmental decision making such as cross-media pollution control and ecosystem management, promotion of large-scale environmental restoration projects, adoption of full cost accounting mechanisms to better measure the value of environmental impacts, and expansion of government support for development of new environmental technologies.[3]

A New Culture at the EPA?

The Bush and Clinton presidencies brought important changes in the use of new approaches to environmental policy. They have been most noticeable at the EPA, where the agency has been shifting away from heavily prescriptive environmental regulation. It now actively seeks flexible mechanisms for achieving environmental goals, reduction in regulatory burdens and costs, experimental partnerships with industry, and voluntary action by industry and other target groups. The agency has begun to use ecosystem management and ecological risk assessments in its water quality, and other, programs. Environmentalists have not been happy about all those changes, and they remain skeptical about how much the use of some of those new policy approaches can alter agency priorities given statutory constraints and political pressures on policy makers. They maintain a similar suspicion about the extent to which the "greening of industry" will modify corporate behavior.

The terms of environmental policy debate, however, have changed. After 25 years of conflict between environmentalists and the business community, the Bush and Clinton administrations tried to build a new consensus through these kinds of reforms. Bush established a President's Commission on Environmental Quality that brought together business and environmental leaders to recommend private sector actions that could promote sustainable development. At the beginning of the Clinton administration, the independent and bipartisan National Commission on the Environment, with solid representation by former EPA administrators, urged a national focus on sustainable development to underscore the inextricable linkage of environmental and economic goals. It endorsed a series of policy and institutional changes consistent with those summarized above (National Commission on the Environment 1993). The National Performance Review (1993b) issued by Vice President Al Gore did the same.[4]

The Clinton White House has been especially eager to move beyond confrontation and policy gridlock. An aggressive environmental backlash in Congress and the nation gave it little choice if environmental policy goals were to be safeguarded. Still, the president was predisposed to favor moderate solutions and to accommodate the critics. Clinton administration officials have used consensus-building approaches to good effect in several cases over the past year, most notably in fashioning Superfund renewal proposals through an ad hoc task force of business leaders and environmentalists. Other attempts to find an acceptable middle course have been less successful, bringing criticism from one side or the other, and sometimes both. That was the case with the Clinton forest plan for the Pacific Northwest.

The pragmatic style was particularly evident at the EPA, where Carol Browner continued initiatives begun under Bush administration EPA head William Reilly to explore the potential of nonregulatory policy approaches. A year and a half after leaving office, Reilly reflected positively on the Clinton EPA: "The culture of the agency is more mature and sophisticated than it has ever been," he said. "The respect for science and for economics is more profound" (Kriz 1994f, 1466). The science may soon get better, as it must to improve the acceptability of EPA regulatory programs. Browner announced in July 1994 that she would seek greater oversight and coordination of the EPA's 12 research labs through creation of four "mega-labs." They are to be devoted to basic research on health and environmental effects, monitoring of exposures to environmental hazards, risk assessment, and pollution prevention and remediation technologies (Stone 1994b). The plan is part of the agency's response to criticism that it has lacked a strong scientific basis for much of its regulatory effort. The EPA is committed to spending at least half of its research budget on long-term projects, expanding peer review of its work, and funding more extramural research. All are expected to foster greater credibility in the agency's science.

Reorganizing the EPA's science labs does little, of course, to overcome other institutional impediments to effective environmental decision making at the agency and throughout the federal government. The EPA still needs to improve its implementation of environmental statutes, use its budget more efficiently, boost staff morale, and overcome fragmentation in environmental regulation. As discussed in Chapter 4, it is unlikely to make much progress in these areas without considerable assistance from the U.S. Congress in restoring enough discretion to the agency to permit intelligent planning and priority setting. The agency's Common Sense Initiative, begun in July 1994, offers some promise of coordinating regulations across air, water, and land media. It may indicate the beginning of an era of greater rationality in environmental policy at the EPA.

Environmental Equity

Recent criticism of environmental policy on equity grounds will continue to generate issues likely to be high on the agenda for the rest of the 1990s. Environmental justice groups have faulted public policy, and the EPA in particular, for insufficient attention to the inequitable burden placed on poor and minority communities that have higher rates of exposure to toxic chemicals and other pollutants (Bryant and Mohai 1992). The EPA's own elaborate 1992 study confirmed a disproportionate impact on poor and minority groups (U.S. Environmental Protection Agency 1992b). It urged an increased priority within the agency to equity issues and to targeting of high concentrations of risk in specific population subgroups. It also highlighted the need to collect environmental data in a manner that allows assessment of risk by race and income and to change EPA risk assessment procedures to better characterize risk across populations and geographical areas.

Efforts to institute these recommendations were underway in 1994, assisted by the EPA Office of Environmental Justice (OEJ). Created in 1992 as the Office of Environmental Equity, the OEJ is a centralized EPA office dealing with environmental impacts on minority and low-income populations. It serves as a clearinghouse for distributing equity information to EPA staff and the public and it helps to coordinate pertinent agency programs. One of its major functions is to assist communities in environmental cleanup activities.[5]

On February 11, 1994, President Clinton issued Executive Order 12898, which called for all federal agencies to develop within one year strategies for achieving environmental justice. The EO established a White House interagency working group to guide the agencies and coordinate these efforts. The order also reinforced existing law by forbidding discrimination in all agency policies, programs, and activities on the basis of race, color, or national origin.[6] In addition to these administrative initiatives, several environmental justice bills introduced in Congress sought to achieve many of the same objectives of nondiscriminatory compliance with federal environmental, health, and safety statutes (Camia 1993b).

Interest in questions of environmental equity has grown rapidly over the past several years within and outside of government. In part, the new visibility of these issues reflects successful grassroots organizing in poor and minority communities. It also signals a new willingness in established civil rights organizations to devote more attention to environmental health issues, and a determination in the national environmental organizations to incorporate environmental justice concerns.

The trend is important for another reason. More than any other recent development, environmental equity concerns force policy makers to confront the ethical issues inherent in choosing environmental policy strategies. Whether the focus is on inequitable exposure to environmental pollution across population subgroups within the United States or other nations, or the far greater economic and environmental inequities between poor and rich nations, the contrasts are striking. They make clear that policy decisions involve more than questions of environmental science and technology. They also tell us that consideration of economic impacts cannot be confined to aggregate national costs and benefits. The distribution of risks, costs, and benefits across the population, and across generations, must be more directly assessed in evaluating environmental policy.

Industrialized nations have recognized those needs in setting new directions in international environmental policy. For example, in 1994, representatives of 64 countries agreed to ban the shipment of hazardous wastes from industrialized to developing nations; the ban is to be fully in effect by 1997. The shipment of wastes had become symbolic of inequitable treatment of poor nations.

Improving Institutional Performance

Another environmental policy need, as reiterated throughout the book, is improvement in governmental performance. This can be done, for example, through ensuring adequate agency budgets and professional staff and reorganizing agencies, creating new capabilities for policy making and implementation where needed. Most agencies have been chronically underfunded and hence unable fully to achieve policy goals even when officials committed to policies and programs are put in charge. Unfortunately, one of the long-term legacies of the Reagan era is a government-wide fiscal austerity that will continue for the indefinite future to constrain programs. Congressional efforts in 1995 to require a balanced federal budget while simultaneously cutting taxes and protecting most entitlement programs could result in a new round of sharp cuts in environmental programs. These conditions force agencies, as they do colleges and universities in the 1990s, to think of new ways to meet goals through institutional reforms and priority setting.

For environmental agencies, reorganization presents at least some opportunities to build a capacity for improved management while also promoting efficiencies. Secretary of Energy Hazel O'Leary tried to reform the Department of Energy's waste management and environmental restoration programs in part through revisions in departmental contracting procedures that will eliminate enormous waste. Secretary of Interior Bruce Babbitt's creation of the National Biological Service and other actions at Interior promised to reverse long-dated and inefficient policies while strengthening capacities for ecosystem management.

The Clinton administration's National Performance Review (1993a,b), like executive studies before it, identified hundreds of ways to improve federal management. State and local governments, as illustrated in earlier chapters, often have been at the forefront of environmental policy innovation, which has included novel institutional initiatives (John 1994; Lowry 1992; Rabe 1994). The trick in these cases is to design appropriate policies and other reforms, devise a strategy for overcoming institutional inertia, and build a constituency for improved governance through public education and political leadership. The successful cases show that it is possible to reform government. Few would contend, however, that it is easy to do so, or that such reform by itself is sufficient to achieve long-term environmental goals.

The Clinton Presidency: High Expectations and Harsh Judgments

The Clinton administration illustrates the pitfalls as well as the promise of environmental reforms of this kind. A president's capacity to lead a fragmented and fractious U.S. political system has declined appreciably in recent years. The public and the press, however, continue to hold presidents responsible for the performance of government, including environmental policy actions and achievements. The environmental community had great expectations for policy change following 12 years of the Reagan and Bush administrations. Those expectations were heightened by the election of the Senate's leading environmental advocate, Al Gore, to the vice presidency.

Clinton nominated leading environmentalists to high positions in his administration, most notably Bruce Babbitt, and dozens more within the Department of the Interior, EPA, and other agencies. Babbitt had been president of the League of Conservation Voters just prior to his selection as Interior secretary. The president selected Brooks Yeager, the former top lobbyist for the Audubon Society, to head the Office of Program Analysis at Interior. He named George Frampton Jr., president of the Wilderness Society, an assistant secretary at Interior with responsibilities for overseeing the National Park Service and Fish and Wildlife Service. The former legislative director of the Sierra Club's Washington office, David Gardiner, became EPA assistant administrator for policy, planning, and evaluation.[7]

Despite the president's initial good relations with the environmental community, at the end of his first year in office, environmentalists rated Clinton a poor achiever. Greenpeace awarded Clinton an "A for Rhetoric; D for performance." The League of Conservation Voters gave him a C+ for overall actions, an A on appointments, a D+ on budgets, and a C− on delivery.[8] Many groups, including the Sierra Club, expressed bitter disappointment with Clinton's support for the North American Free Trade Agreement. Others complained of his failure to strongly back Bruce Babbitt's natural resource reform agenda. Clinton, they said, was willing to compromise far too quickly on public lands and other issues, and he failed to use the still considerable resources of the presidency to advance environmental causes.

Even the *New York Times* expressed disenchantment with the president's environmental leadership, although the editors correctly observed that any blame for inadequate environmental policy had to be shared by members of Congress. The editors at the *Times* accused the president of a lack of nerve in confronting fellow Democrats in Congress who, among other actions, voted against the long-disputed EPA cabinet bill over allegations of environmental policy overkill: "Mr. Clinton has the chance to become the 'environmental President' George Bush never really became. But selling his agenda to a contentious Congress will take a lot more dedication than he has so far displayed" (New York Times 1994a).

Like the Bush White House, the Clinton presidency receives fewer accolades than its environmental initiatives warrant. High expectations and disappointing results partially explain the phenomenon. So does the unwillingness of presidents to talk plainly with the American public about the environmental problems the nation faces and the tough choices to be made. Regrettably, the current state of U.S. politics does not encourage the public dialogue that is so clearly needed.

TOWARD SUSTAINABLE DEVELOPMENT

The political process also pushes policy makers toward preoccupation with short-term policy disputes and the tasks of reconciling diverse interests. Even when consensus exists on long-term goals such as sustainable development, they find it difficult to get a firm fix on how that abstract concept affects short-term decisions on controversial public policies. It does not help that the government, like most other organizations, lacks the capacity for long-term foresight and policy planning. Instead, we muddle through, making incremental adjustments as needed. The strategy may work well when the rate of change is slow enough to permit such adaptation. Higher rates of change are problematic for governments as well as other organizations. This is especially so when decision makers must act without sufficient knowledge of the scope of change, the probable impacts on the environment and society, and the effectiveness of alternative strategies of response.

The United States and other nations find themselves in precisely this predicament as they move toward the twenty-first century. Forecasts of changing environmental conditions offer little basis for complacency. Yet they also come with enough uncertainty that disputes over the facts and the logic of competing policy strategies prevent agreement and action. We desperately need to develop the institutional and political capacity to improve scientific forecasts as well as to foster the public consensus essential to promote sustainable development.

This expansive environmental agenda for the twenty-first century can be illustrated with actions the Clinton administration is taking to foster political discourse on sustainable development; continuing needs for policy formulation in the areas of energy, population, biodiversity, and environmental research; and the formidable task of implementing the Earth Summit's Agenda 21.

The President's Council on Sustainable Development

Bill Clinton's creation in mid-1993 of a President's Council on Sustainable Development (PCSD) suggests what might be done to tie short-term policy decisions to long-term environmental goals, even if the council's impact has yet to be demonstrated. From the Brundtland report of 1987, *Our Common Future,* to the Earth Summit's Agenda 21, environmentalists have pressed government to make sustainable development the central guiding principle of environmental and economic policy making. The PCSD is setting the foundation for such a change in policy perspectives and institutional practices. It even adopted the Brundtland Commission's definition of sustainable development: "meeting the needs of the present without compromising the ability of future generations to meet their own needs."

To date, the council has functioned as an "envisioning" body. It brings together top corporate officials, leaders of environmental organizations, and administration officials, including members of the cabinet, to explore the connection between the long-term environmental agenda and current policies. In some respects, it is the domestic equivalent of the United Nation's Commission on Sustainable Development, which grew out of the 1992 Earth Summit.

The council was given two years (until July 1995) to define long-term environmental goals that can help to integrate U.S. public policies. It has a small staff of 15 and a modest budget of $1.5 million, which ordinarily would not encourage a positive assessment of its political influence. Its reach appears to be greater than those numbers suggest. The council operates primarily through a series of task forces dealing with subjects such as ecoefficiency, sustainable communities, ecosystem management, bioregionalism, integration of energy and transportation systems, and full cost accounting. Because of the involvement of cabinet members, the council is shaping policy formulation within the agencies long before it writes its final report. For example, Secretary of Energy O'Leary credits the council's energy task force for contributing to the DOE's development of the Clinton administration's National Energy Policy. Some accounts of the council's activities hint at comparable influence on policy maker attitudes and policy formulation (Russell 1994).

The PCSD's activities suggest the potential for comparable action by state and local governments. In cooperation with industry and environmental organizations, they too need to formulate strategies for sustainable development. Some of them already are doing so as illustrated by California's development of a long-term energy policy. State and local governments will play an especially crucial role in effective land-use planning, a function historically handled at the local level. With encouragement and support from the federal government, states and localities can fashion land-use planning policies that promise to protect environmentally sensitive areas such as wetlands and watersheds as well as biological diversity, encourage energy conservation and efficiency, sustain the productivity of agricultural land, select appropriate sites for locating industry and housing, and meet the many other environmental needs that fall within local jurisdictions. Public involvement in the formulation of such land-use plans should help to ensure responsiveness to public needs and the requisite degree of public support for implementation of the plans.

New Policy Horizons

As the PCSD's work demonstrates, sustainable development requires that the nation deal with several fundamental activities largely beyond the reach of current public policies. These include the level of energy use and the energy sources relied upon, the size and growth rate of the population and its habits of consumption, and the prevailing use of technologies—from transportation to agriculture. All must be consistent with the maintenance of ecosystem health on which life depends, and decisions in each area will be affected by the character of scientific knowledge and how it is used.

Federal, state, and local policies affect energy use, population rates, patterns of economic development, and ecological systems to varying degrees, mainly through indirect effects on private decisions. At the global level, bilateral treaties, regional associations, and international agencies such as the United Nations Environment Programme and the World Bank have comparable effects. More explicit and direct policies have been proposed at all levels, and many more will emerge in the years ahead. Policy entrepreneurs will continue to develop, incubate, and refine them, and to test them against the prevailing political winds. Several of those proposals merit brief review here.

Energy Policy

As the review of energy issues in Chapter 5 noted, the United States has no comprehensive national energy policy. The nation has made substantial advances in energy efficiency and conservation, and in furthering the use of renewable energy sources. Government policies have helped to stimulate those changes, and the Clinton administration promises increased support for such new policy directions. Nonetheless, the United States remains dangerously dependent on fossil fuels for its energy supply, and on imported oil in particular.

A long-term and environmentally sustainable energy policy would curtail those dependencies and foster renewable energy and reduced demand through conservation and efficiency. A policy that meets global standards for equitable use of energy resources (and minimization of greenhouse gas emissions) would aim to reduce U.S. energy consumption well below current levels. Far more is possible along these lines than has been done to date, as some state and local energy policies and recent policy analyses demonstrate (Flavin and Lenssen 1994).

Among other energy policy needs is the development of a safe and publicly acceptable method for disposal of high-level wastes from nuclear power plants. The United States and other nations continue to struggle with those issues without clear resolution (Dunlap, Kraft, and Rosa 1993; Shrader-Frechette 1993). Secretary O'Leary has authorized a DOE review of U.S. nuclear waste policy and members of Congress have called on President Clinton to establish a presidential commission to study the controversial program. Concern about greenhouse gas emissions from fossil fuels has led to new interest in the potential contribution of nuclear power to the world's energy mix. Yet without a solution to the nuclear waste issue, expansion of nuclear power is doubtful.

During 1994, the Clinton administration was developing its own version of a national energy policy for submission to Congress. If the president's Climate Change Action Plan announced in late 1993 is any guide, the new policy proposals are unlikely to break new ground. The climate change strategy relies largely on voluntary measures to cut greenhouse gas emissions (favored by the business community) and it avoided the more aggressive action favored by environmentalists and recommended by many scientific studies.

Cutting such emissions to 1990 levels by the year 2000 is a good first step, but it is not sufficient, particularly in light of expected worldwide growth in population and fossil fuel use (Stevens 1994b). To date, neither the president nor Congress appears ready to consider increases in federal energy or gasoline taxes large enough to reduce consumer demand. Congress also has displayed little interest in raising auto fuel-efficiency standards. Clinton's failure to speak out more forcefully on climate change and other energy issues does not auger well for U.S. leadership on the U.N. Convention on Climate Change and related measures.

Population Policy

There are few policy areas that were more seriously neglected in the 1980s than population. The Reagan and Bush administrations took the position that the nation needed no policy. To appease important conservative constituencies, they ended all U.S. contributions to the United Nations Fund for Population Activities (UNFPA), later renamed the United Nations Population Fund. The UNFPA had been (and remains) the major multilateral funding agency for international family planning programs. The fund cutoff followed a 1984 UN conference on population in Mexico City at which the United States announced its new policy on population.

Although the elimination of U.S. financial support for UN population programs lasted from 1986 to 1992, Presidents Reagan and Bush did maintain support for U.S. bilateral population assistance programs, where the nation directly supplies funds to other countries. By fiscal 1993, those programs received more than $400 million a year, largely to support family planning programs in developing nations. Nevertheless, the shift in population policy in the 1980s contributed significantly to the virtual disappearance of population issues on the national agenda. Failure to address population growth and its environmental consequences was astonishingly shortsighted given U.N. projections of a probable doubling in the world's population over the next 60 years.

The Clinton administration quickly reversed the so-called Mexico City Policy, and it reestablished support for the UN population program. It also boosted funding for the nation's bilateral family planning assistance programs, proposing more than $580 million a year in fiscal 1995. The administration already had signaled that it would once again exert a leadership position on world population issues. It did so at the 1994 UN Conference on Population and Development in Cairo, where delegates debated a proposed action plan on global population stabilization.

Most of the controversy at Cairo focused on the abortion issue, with the Vatican and many Muslim and Latin American nations objecting strenuously to acknowledgment of legal abortion being provided as part of reproductive health care. Yet despite those disagreements, delegates to the historic meeting, including those representing the Vatican, demonstrated a remarkable consensus on the basic principles of population policy for the twenty-first century. Representing 180 nations, they approved a 20-year "program of action" that they believed could hold world population to around 7.8 billion by the year 2050. The 113-page declaration called for a tripling of the amount the world spends on population stabilization, rising from about $5 billion spent in 1994 to $17 billion by the year 2000. Delegates also expanded the concept of population policy well beyond traditional family planning programs. The action plan calls for governments and donor nations to support education for girls to help promote gender equity, to provide women with a range of choices on family planning and health care, and to improve the status of women in developing nations so they are empowered to make decisions related to reproductive choice (Cowell 1994).

One of the most notable omissions in U.S. population policy is a failure to set any national goals for the United States itself (Kraft 1994b). The latest Census Bureau projections put the U.S. population at some 390 million by the year 2050, an increase of almost 50 percent over the present population size (Population Reference Bureau 1994). Federal policy makers are reluctant to deal with the issues. Yet a larger population increases demands for homes, automobiles, and other consumer products, and the energy and natural resources their production and use require. It also raises serious questions of equity in the use of the world's resources and whether the environment's carrying capacity will be exceeded (Mazur 1994; Postel 1994).

At any given time and place, prevailing technologies as well as social values affect the carrying capacity of the environment. In the face of rapid population growth and continued high per capita consumption of material goods, sustained economic growth depends upon unpredictable breakthroughs in information and technological systems and/or a reallocation of critical resources such as land and water to more efficient uses. Prudent policy makers would choose to act sooner rather than later to promote sustainability that ensures a high quality of life and maintenance of cherished social values. By delaying action and avoiding the hard decisions, they will have fewer and less attractive options in the future.

Maintaining Ecosystem Health and Restoring Biodiversity

Any program of sustainable development must include policies to halt the loss of biodiversity, restore ecosystem health where damaged, and maintain the ecological functions essential to long-term environmental sustainability. Controversies over the Endangered Species Act attest to the obstacles that lie ahead in formulating national and international programs to achieve those goals. The Clinton administration recommended revisions to the ESA that would encourage a proactive strategy of protecting threatened ecosystems rather than the ineffective approach of identifying and trying to save single endangered species or setting aside small nature reserves. Scientists have long recommended this kind of break from outmoded conservation paradigms. Recognizing the need to find a politically acceptable course of action, Clinton also called for making any such strategy sensitive to human needs and the rights of property owners. The policy's success will depend on how policy makers strike such balances.

Much the same is true of international biodiversity programs that seek to integrate economic development with habitat preservation. The Convention on Biological Diversity, a decade in the making, took effect in 1994. Its goal is to slow the loss of the Earth's biodiversity through adoption of national policies to conserve species and their habitats and to promote public awareness of conservation and sustainable uses of biological diversity. An international meeting held late in 1994 developed implementation guidelines for the treaty.

Some comprehensive efforts at ecosystem restoration, such as the remedial action plans in the Great Lakes Basin and the Florida Everglades, have been underway for several years. Others, such as the planned recovery of the oak savanna in the Midwest and the coastal sage ecosystem in California, are just beginning. Some environmentalists urge far-reaching actions that dwarf present activities. For example, the Society for Conservation Biology has revealed a Wildlands Project that envisions setting aside vast tracts of land for wildlife reserves to ensure the survival of native species, particularly large carnivores such as wolves, grizzly bears, and mountain lions; recent scientific studies indicate they need exceptionally large areas for a viable habitat. The project calls for over 23 percent of land on the Oregon coast to be returned to wilderness, and another 26 percent to be highly restricted for human use. Similar plans were developed for Vermont, Florida, the mid-Atlantic region, and the rest of the nation. One of the plan's authors, ecologist Reed F. Noss, described it as "a vision of what this continent might look like in 200 years if we can reduce the scale of human activities" (Mann and Plummer 1993). Some ecologists call the plan a logical extension of the ESA. However, the groups and policy makers already highly critical of the present law would doubtless be less enthusiastic about such an expansion of its scope.

Even efforts well short of the Wildlands proposal will test the limits of public and governmental support for preserving biodiversity. If recent experience is much of a guide, setting aside large areas for biological reserves will work only if the public becomes far more knowledgeable about the critical role of biodiversity in any scheme of sustainable development. Stakeholders, including local residents, also must be involved in goal-setting and management decisions from the earliest stages of any program of biodiversity protection. They must share in the vision of sustainable management of biodiversity resources and feel responsible for maintaining a commitment to long-term goals. No one would argue this would be a simple undertaking. Without it, however, short-term human needs and desires are likely to overwhelm plans for long-term environmental preservation.

Environmental Science and Policy Research

Learning more about the environment and human interaction with it is crucial to policy success, but environmental research has been poorly funded and cannot meet the demands placed on it. Total federal environmental research runs to about $5 billion a year, of which 10 percent finds its way to the EPA. The rest is divided among 20 other federal agencies. Even with the higher levels of support over the last few years, the scientific community argues for increasing the amounts available for environmental science and policy research (Carnegie Commission 1992, 1993a).

Present spending on environmental science and policy research still pales in comparison to the government's overall $75 billion R&D budget. It is especially small in relation to spending on military research, which has consumed more than half of that amount. As military threats to the nation's security diminish, a strong case can be made that the concept of national security must be expanded to incorporate threats posed by climate change, depletion of the ozone layer, and similar environmental risks (Funke 1994). That is, in the 1990s and twenty-first century, environmental hazards may pose greater risks to the nation's long-term security than those related to military actions. Changing the pattern of research spending is part of that new equation.

The federal government has recognized some of these needs, but agencies and scientists jealously guard their research funds. As noted, the EPA has announced its intention to reorganize the agency's elaborate research laboratory system. Similar shifts have been underway with military research laboratories as part of what is termed "defense conversion" following the end of the Cold War. For example, in 1990, the departments of Energy and Defense jointly established a Strategic Environmental Research Defense Program geared to the problems of cleaning up the environmental legacy of the Cold War. In a move that could result in reallocation of federal research dollars, in late 1993, President Clinton established by executive order a new National Science and Technology Council. It operates through ten panels overseeing the major R&D programs, and includes a Committee on Environment and Natural Resources. The council also seeks to develop a long-term policy on federal scientific research. The first evidence of the council's influence should appear in the fiscal 1996 budget.

For most environmental policies, the key issues of environmental research involve the acquisition and use of data to improve program operations more than they do pursuit of basic science. Both are essential, but the former is the more pressing need at present. It is a central element in proposals for a National Institute for the Environment (NIE). The Committee for the NIE, headed by former Ambassador Richard Benedick, argues that the institute is necessary to fund credible, problem-focused interdisciplinary and multidisciplinary scientific research. Sponsors believe creation of the NIE will help ensure thorough assessments of environmental knowledge and identification of gaps in present knowledge and research needs. The NIE also should help to expand and facilitate citizen access to environmental information through a computer-based National Library of the Environment and to improve environmental literacy through support for environmental education and training programs in public schools and universities (Howe 1993).[9]

The assessment of knowledge and making it available to the public and policy makers is particularly important if environmental decision making is to be more firmly based in reliable science. So too is the intention to encourage and support interdisciplinary and multidisciplinary research that is essential for understanding and responding to complex and multifaceted environmental problems such as global climate change and loss of biological diversity. The NIE proposal continues to undergo review and debate, and its approval is by no means certain.

The NIE or an institution much like it is needed to supplement the often narrow, fragmented, and short-term focus of agency-funded scientific research, which too often is done in reaction to particular program needs and without peer review. The effectiveness of environmental policy depends on broader, coordinated, long-term, and problem-focused research. Such research also could draw far more from the underutilized potential of the nation's scientific community outside of government. A recent study by the National Research Council highlighted the present deficiencies in the nation's research programs. It found no strong commitment to environmental research at the highest levels of government in comparison to research on the military, health, and transportation, nor any coordinated national plan for environmental research. What research there is focuses almost exclusively on the natural sciences, with little attention to the impact of environmental changes on society. The report also found weak linkages among science, policy, and management in federal environmental R&D.[10]

Consistent with these criticisms, the EPA's Science Advisory Board recommended late in 1994 that the agency develop methods to foresee impending environmental problems to facilitate early action on them. It hoped that the EPA would begin focusing more on environmental problems of the future and less on those of the past and present. The SAB urged the agency to begin issuing an annual report describing likely environmental conditions 20 years in the future.

International Environmental Policy

Among the greatest policy needs for the twenty-first century is enhancement of international environmental efforts. A large body of international law established over the past several decades helps to govern the global environment, assisted by the policies and programs of international agencies such as the UN Environment Programme (UNEP) and the World Bank (Caldwell 1990; Soroos 1994; Porter and Brown 1991). The 1992 Conference on Environment and Development (UNCED) and many others held since 1970 on food, population, water, human settlements, and climate defined the issues and spurred international agreements. Taken together, these activities have accomplished much (Starke 1990). However, any assessment of additional requirements to protect the global environment and to assist nations in moving along a path of sustainable development must be grounded in political realities that are less encouraging.

Institutions such as UNEP and the conference diplomacy that has characterized international environmental policy are inherently limited in bringing about significant policy and institutional changes, particularly in the short term. UNEP, for example, has been a small, poorly funded and staffed agency with little operational authority. It has served primarily as a catalyst for environmental action by member states. International treaties and agreements such as the Convention on the Law of the Sea (UNEP counted more than 150 of them through 1990) can have substantial effects. Yet the coverage of such laws is uneven and they apply only to nations that agree voluntarily to comply with them. No international institution has legal authority over nation states comparable to their own internal governments, and each nation continues to define its interest based on concepts of national sovereignty. Moreover, international agreements often are not well monitored and enforced, and in many cases their specifications are too vague to permit evaluation of their achievements.

To illustrate the general problem, environmentalists fault the World Bank and other lending institutions for their continued emphasis on large, environmentally destructive development projects despite years of assurances that they will assess environmental impacts of their projects and alter priorities (French 1994). Brazil continues to resist what it considers to be outside interference with its management of the Amazon forest, and it has encouraged rapid deforestation in an effort to speed economic development.

These institutional and political obstacles seriously impede the ability of international institutions to bring about the kinds of changes incorporated into UNCED's Río Declaration and Agenda 21 (United Nations 1993). The former is a statement of

27 general principles on human needs, environmental sustainability, and the responsibilities and rights of nation states. Agenda 21 is a formidable collection of policy goals for the next century intended to move the world toward environmental sustainability while simultaneously providing essential economic development. The 1987 Brundtland Commission report (World Commission 1987) shaped the intellectual framework for both.

The UNCED secretariat has estimated that the cost of implementing Agenda 21 will be about $600 billion per year from 1993 to 2000. Developing nations would provide most of those funds. However, the plan calls for the industrialized nations to contribute $125 billion a year beyond what they already provide under a variety of existing programs. The target aid level is 0.7 percent of GNP "as soon as possible." In recent years, U.S. aid has totaled about 0.2 percent of GNP, or less than amounts contributed by 14 other nations. The new spending, part of which is to be funneled through the Global Environment Facility (GEF), would roughly double current levels of support (Jordan 1994a). The GEF is an international financing mechanism originally established as a partnership involving the World Bank, the UN Development Programme, and UNEP to provide developing nations with the means to bypass polluting and wasteful technologies and move toward sustainable economic development. It was restructured after UNCED as a larger and better financed agency to assist in implementing international conventions on ozone depletion, biodiversity, and climate change as well as provisions of Agenda 21 (Jordan 1994b).

The first report of the panel set up to monitor achievements following the Río conference hinted at implementation problems ahead. In May 1994, the UN Commission on Sustainable Development indicated that only slow progress was being made. The commission chair said the international community had fallen "significantly short of expectations and requirements" in providing money and the technological expertise developing nations need to foster environmental sustainability. He called on the rich nations to do more to restrict their own environmentally harmful consumption patterns (particularly energy use). However, the commission made no effort to evaluate the individual reports submitted by nations on their own progress as required under the Río agreements. The chair also acknowledged that he and other environment ministers could accomplish little when more powerful departments in government sought to promote industry, exports, and jobs (P. Lewis 1994).

Whether the present ensemble of international institutions, agreements, and agendas succeed in reaching the goal of environmental sustainability will depend on public support, governmental commitment, and especially political leadership within each nation. There are good reasons to be skeptical that all will be forthcoming regardless of the urgency of action. Within the fragmented U.S. political system, for example, such leadership can come only from the White House. Yet it was largely absent under Ronald Reagan and George Bush, and it has been less evident under Bill Clinton than environmentalists had hoped to see. The Clinton administration has been supportive of the UNCED climate change and biodiversity agreements, and it has taken a strong stance on the need to reduce world population growth. Nonetheless, questions remain about the extent of the administration's commitment to long-term environmental goals and the priority the White House is likely to give to such issues.

These disappointments notwithstanding, the goals and the processes that UNCED set in motion remain highly important. As UNCED organizer Maurice Strong observed, the 1992 meeting was a "launching pad," not a quick fix. The previously fuzzy concept of sustainable development was given a clearer form and guiding principles and goals were set in place that will shape economic and environmental decision making over the next several decades. The mutual dependency of environmental health and economic well-being was firmly established, and realization of such relationships set a new context for international politics and ethics in the twenty-first century.[11]

One basis for modest optimism can be found in the Gallup Health of the Planet Survey conducted in 1992. As noted earlier, it found that in nearly all nations of the world, majorities displayed a high level of environmental concern and, should a conflict exist, favored environmental protection over economic growth (Dunlap, Gallup, and Gallup 1993). If such public concern and support for policy action continues, it should form a solid foundation for the exercise of political leadership on behalf of environmental sustainability.

BUSINESS AND THE ENVIRONMENT

No matter how carefully designed and effective public policies are, governmental activities are inherently limited in their ability to change consumer behavior and industrial activities. Thus environmental sustainability depends as well on actions taken in the private sector, including those by business organizations. That realization both cheers and worries environmentalists.

The Greening of Industry

Business groups often have been active opponents of environmental protection policy, and pursuit of profit by private corporations and landowners is responsible for much of the abuse of natural resources in the United States and globally. Many leading business groups continue their efforts to weaken federal environmental laws they believe are too costly. Yet some of the largest and best known U.S. corporations have demonstrated a new willingness to foster sustainable resource use, to support pollution prevention initiatives, and to develop and market green products.

The greening of industry is evident in many quarters, stimulated by federal environmental policies and changing market conditions. Pollution prevention as a strategy has proved to make good economic sense for many corporations. They save money by reducing or eliminating the production of wastes and pollutants rather than by disposing of them or cleaning them up afterward. They use cleaner technologies, improved production processes, better controls and materials handling, and materials substitution (Hirschhorn and Oldenburg 1991). Federal actions such as publication of the annual TRI report on toxic chemical releases provide incentives for companies to minimize pollution and avoid public censure. Consumer demand for environmentally benign and energy-efficient products also has been important, although it is uneven and often insufficient to induce corporations to develop and market new products they fear will find too few buyers.

Some of the most revealing indications of a change in outlook in the business community came in *Changing Course* by Stephan Schmidheiny and the international Business Council for Sustainable Development (1992), released just before the 1992 Earth Summit. The council, consisting of 48 industrialists from companies such as Chevron, Alcoa, Du Pont, Dow Chemical, and Johnson Wax, outlined a business commitment to sustainable development. It also offered plentiful examples of how open and competitive markets can foster innovation and efficiency and provide opportunities for sustainability initiatives. This potential depends heavily on whether markets give the right signals with prices of goods and services that reflect environmental and resource impacts.

Recent governmental studies (e.g., U.S. Office of Technology Assessment 1992) similarly urge the development of cleaner products through better design and life-cycle analysis and product stewardship, sustainable management of renewable resources in agriculture and forestry, and similar actions. Such studies conclude there is much potential as well in the transfer of environmental technologies to developing nations. Many of the nation's leading corporations, including Minnesota, Mining, and Manufacturing (3M), Du Pont, Monsanto, Shell, Chevron, Johnson Wax, Dow Chemical, Procter and Gamble, and Scott Paper, already have taken steps toward sustainable development. Their actions suggest the potential for future advances. Such shifts in corporate behavior are even more evident among progressive or socially conscious businesses (Schmidheiny 1992; Barnthouse 1995; Hawken 1994).

The continuation and expansion of these efforts depend on public support for them, and particularly on consumer demand. The past several years have brought a profusion of consumer guides to green products and services. Consumers should soon find the task of selecting environmentally appropriate products even easier when new green labeling services become more common and governments agree on the meaning of terms such as "recycled." So-called "product policies," more common in Europe and Japan than in the United States, also can stimulate redesign. Such actions in Germany have led automobile manufacturers such as Mercedes and BMW to build cars that are almost completely recyclable. The U.S. Office of Technology Assessment (1992) has argued that such policies can increase U.S. global competitiveness by reducing production costs, improving quality, and appealing to environmentally sensitive consumers.

A Green Industrial Policy?

Vice President Al Gore and presidential science advisor John H. Gibbons were instrumental in the development of a U.S. "green industrial policy," which they announced in July 1994 in an administration report, *Technology for a Sustainable Future*. The report outlined a strategy to promote the development and adoption of environmental technologies by revising rules that discourage use of new technologies. It also aspires to coordinate green-technology programs in different agencies, institute federal testing and verification of technologies to ensure their performance, assist in the commercialization of new products, and help promote sales of green technologies abroad (Gore 1994). The appeal is understandable with projections of a

$600 billion global environmental technology market by the year 2000 and the potential of shaping a new industrial revolution now in progress (Flavin and Young 1993).

At the state level, studies that focus on the impact of environmental policies on the economy should be encouraging to those who favor such industrial policies. They suggest that states cannot hope to improve their economic performance substantially by reducing environmental standards (Meyer 1993). This is particularly so with a long-term perspective on economic productivity. Hence the business community would be better off by embracing environmental goals and adjusting their operations accordingly.

Environmental protection policies, however, may be especially burdensome on small business with narrow profit margins and a sharply constrained ability to make such adjustments. Maintaining the viability of small companies is necessary for many reasons, including job creation. Those goals could be furthered through policies providing subsidies, loans, or extended deadlines for compliance with new standards. Environmental organizations also have assisted small businesses by organizing pollution prevention workshops and similar educational programs that otherwise would be out of the reach of the smallest companies.

The record of the business community over the past three decades gives environmentalists good reasons to question how far the recent greening of industry is likely to go and to what extent government support of green industrial policies can shift corporate priorities. Actions taken by many corporations over the past several years suggest that fundamental changes are occurring. They can be nurtured through the kinds of reforms discussed in Chapter 6, such as full cost accounting, government purchase of green products, and other policy changes affecting taxation, investment, R&D expenditures, and marketing. Consumers can do much as well. They can exercise their enormous financial power directly in the marketplace as well as through the vast sums they maintain in pension and other retirement accounts. The latter may offer intriguing possibilities. In the last several years, for example, some of the largest state pension funds have successfully moved corporations in the direction of environmental sustainability. Much more could be done if the public and environmental groups brought such pressures to bear on the managers of those funds.

DEMOCRACY AND THE ENVIRONMENT

As much as any other change in society, environmental sustainability depends on public attitudes, values, and behavior. Without a supportive public, governments are unlikely to enact and implement strong environmental policies that are perceived to constrain individuals' lifestyles, limit their rights, or raise their taxes. Nor will businesses market green products that are not otherwise economically defensible. An informed, environmentally committed, and active public provides the incentives policy makers and the business community need to steer a course toward sustainability.

Helping the public to become environmentally literate and to participate effectively in decision-making processes, both public and private, is an essential part of any long-term environmental agenda. Recent trends in U.S. politics offer a sobering picture of the obstacles to democratic participation, but they also hint at the opportunities. Levels of political cynicism have increased over the past 30 years, and they have reached new highs in the 1990s. National surveys during the 1994 election cycle revealed a frustrated and alienated electorate that had little confidence in elected officials or in the political process (Seelye 1994). Such attitudes help to explain some startling shifts in voter behavior in 1994 as many veteran lawmakers were retired and the Republican party captured both the House and Senate for the first time in 40 years. In such a political climate, fostering a constructive public dialogue on environmental issues becomes difficult, particularly at the national level.

Grassroots Environmentalism and Public Education

Despite public cynicism about government and politics, one encouraging sign of the potential for a democratic environmental politics can be found in the rise of grassroots organizations, including those associated with the environmental justice movement in the United States and worldwide (Bullard 1993; Kamieniecki 1993).[12] Another is recent research in political science indicating that the public's capacity for democratic participation has been severely underestimated, even in policy arenas dominated by technically demanding information (Hill 1992; Berry, Portney, and Thomson 1993; Ingram and Smith, 1993). Such research suggests ways in which public involvement can enhance political discourse on technical issues and thereby promote more effective policy choices.

As noted in Chapter 3, opinion polls leave no doubt about public concern for environmental degradation and support for environmental policy. Yet it is equally clear that environmental issues are not highly salient for most people and that even those who are favorably disposed toward environmental values often fail to translate those beliefs into personal action. The public does not follow policy issues closely and it is vulnerable to political rhetoric that exaggerates the effects of environmental policy—for example, its impact on property rights, jobs, and economic prosperity. Environmentalists may compound the problem by treating every ecological and health hazard as a crisis or emergency of the greatest importance. When public fear of environmental risks is greatly at odds with scientific evidence, elected officials may be unable to set realistic priorities. There is also a danger that the public will tire of environmental campaigns, particularly if environmental groups bombard them continuously with the message that fundamental changes are necessary in public policy or in their individual lives without helping them to understand the reasons for such changes.

Part of the solution in these instances is improved public education and involvement (Orr 1992; Milbrath 1989). People need to learn more about the environment in which they live and the risks they face, and they need to enhance their capacities to make sense of diverse assessments of environmental problems. They have to learn how to differentiate between valid and unsound arguments, between self-serving fictions and scientific fact. They need to be able to distinguish environmentally sound products and personal lifestyle choices from those that are wasteful and damaging. People especially need to be able to relate their perceptions and judgments about their immediate environments, the places in which they live and work, to the more abstract environmental arguments they hear at the state, national, and global levels.

Environmental groups and individual activists act as surrogates for the larger public. That is an essential need in a large and complex society. Yet those groups function best in cooperation with an informed citizenry that is prepared to become more directly involved, particularly at the local level. In the last several years, government agencies also have recognized that they cannot hope to secure public approval for many activities, such as the siting of waste facilities, without providing sufficient opportunities for citizens to participate in critical judgments about acceptable levels of risk. Participation without knowledge is risky because people may make poor choices and resources may be wasted. But decision making without public involvement invites ethical lapses and policy failure (Dunlap, Kraft, and Rosa 1993; Slovic 1993; Rabe 1994). Democracy requires citizen interaction and deliberation, and we need to continue to search for the ways in which those political values can best be realized.

The Risks

Two trends affecting public participation in environmental decision making are especially encouraging. One concerns generational differences in environmental attitudes and behavior. The other has to do with technology, and particularly that associated with the much vaunted information superhighway.

Surveys of environmental attitudes reveal few significant demographic differences. A persistent, if minor, one has been age. Younger segments of the population evince a greater concern about the environment than do other groups. That finding is reinforced by a nationwide Peter Hart survey in 1992 that showed for Americans under age 18, "protecting the environment" was the "single most important issue for America to work on in the future," far outranking concern over drugs, education, homelessness, and the economy (Brody 1993). Further evidence is found in the phenomenal success of popular books such as *50 Simple Things Kids Can Do to Save the Earth* and a sequel, *Kid Heros of the Environment.*[13]

This interest has kindled the development of hundreds of environmental groups as well as newsletters and instructional programs on the environment that are geared to children, including a multitude of television programs and cartoon shows. Some of the same forces affect product development and sales. A Rand Youth Poll in the early 1990s found teenagers have a $230 billion a year impact on the U.S. economy. Realization of the potential of this spending has contributed to the expansion of green marketing. New green products have been appearing at 20 times the rate of other new packaged goods, with many of them targeted at children (Better 1992). If these attitudes persist as the present generation ages, environmental protection activities should enjoy an even stronger public backing in the future.

Computers and Environmental Activism

One development affecting the way environmentalists, youthful or otherwise, will be able to inform themselves, organize, and respond quickly to changing issues is the spread of computer technology. In mid-1994, the Intel Corporation estimated that 30 percent of U.S. households had a personal computer, and it forecast a rise to 50 percent by the year 2000. The rate is much higher for affluent families. Widespread use of computers is driving an explosive development of green software, databases, networks, and bulletin boards, the full effects of which are unclear but promising.

Government agencies are supplying much of the data moving through those networks. For example, the Right-to-Know Computer Network (RTK Net), operated by two Washington, D.C., nonprofit groups, offers users online access to the TRI database. Congress helped by insisting that the TRI data be released to the public in a computer-readable format. It took the RTK Net to make the data available at a nominal cost to enable community activists to deal with local problems involving toxic chemicals. The TRI database is now available on CD-ROM as well. Other examples of the availability of government data abound. The United Nations has produced a CD-ROM disk containing the full text of all pertinent documents on the 1992 Earth Summit, including many not available in any other form. The EPA now has an extensive array of information available to citizens over the Internet, as do other government agencies.

The amount of environmental information being collected worldwide is staggeringly large and growing rapidly. Much of it, including satellite monitoring and Geographic Information Systems data, will be available to anyone with the necessary software and hardware. Environmental organizations such as the World Resources Institute and Worldwatch Institute, and government agencies such as the World Bank, already provide some of their elaborate environmental data on floppy disks for ready access (Young 1994). Reference sources such as the *Environment Reporter* have begun using a CD-ROM format that should make information searches easier and quicker.

No one can say how much the public will make use of the information superhighway to access and use environmental data of this kind. Yet its availability is an asset that environmental groups and activists are unlikely to ignore. The potential for "eco-linking" conferences and circulation of reports and data across computer networks can only increase over time. One effect will be to lower the cost of acquiring information and to facilitate citizen communication with others having similar interests. Community activists may find it easier and cheaper to exchange views via the computer than to assemble in one physical location. Such meetings already are common on the Association for Progressive Communications network where they help to connect people around the world as easily as they do residents of the same community.

Those who value democracy can help to ensure that data and networks like these contribute to public understanding of difficult environmental problems and to communication about them. Much can be done to assist people to develop the skills they need to put environmental information to good use. None of these developments has yet made conventional governmental and political processes obsolete. They do require, however, that we rethink the way environmental politics will work with the new technology in place.

CONCLUSION

Environmental policy depends on advances on many fronts: scientific research, technological invention, reengineering of industrial processes and improvements in corporate management, enhancement of governmental capacities for policy making and implementation, more widespread embrace of environmental values, and increases in public knowledge and involvement. The late twentieth century is a time of profound and unpredictable changes in these areas of human activity. It is also a difficult period for those who place their hope in governmental and political processes. Critics disparage politics and contribute to public cynicism. One effect is to limit the collective power of citizens and to increase the influence of well-organized special interests that pursue political values at odds with those held by the public. An invigorated democratic politics is essential to any strategy for creating a just and sustainable society, and we need to seek ways to make the political system more open to public participation and more responsive to public needs.

The short-term environmental policy agenda for the 1990s and for the early twenty-first century is clear and broadly supported. There also are signs of progress in achieving it despite continuing battles over environmental protection and natural resource policies in Congress and the states. Individuals will disagree over the interpretation of the latest data and studies, and over the way policy makers should reconcile conflicting positions on the issues. Those who find the direction, form, or pace of policy and institutional changes deficient should consider entering the fray to fight for their views.

The longer-term agenda of sustainable development is less distinct, but it is becoming sharper as societies struggle to define its meaning and shift programs and priorities to promote it. The challenges here are more daunting and demand more commitment than has been evident in either government or corporate circles. Citizens need to press both government and business to do a better job. They can help by building coalitions for environmental sustainability, articulating the social, economic, and political changes necessary, and devoting themselves to their realization.

ENDNOTES

1. Even Vice President Al Gore's *Earth in the Balance* (1992) has provoked a detailed rejoinder by conservative analysts. See John A. Baden, ed., *Environmental Gore: A Constructive Response to Earth in the Balance* (San Francisco: Pacific Research Institute for Public Policy 1994).

2. The committee's report is to be called "Redefining the 100 Basic Assumptions of Mankind," and will be published in book form. The committee's work is described briefly in *World Watch*, June 1994, p. 7.

3. A review of these issues, including recent efforts and remaining tasks, can be found in Vig and Kraft (1994, ch. 17).

4. Two of the accompanying reports of the National Performance Review (1993b) are especially pertinent for the EPA: *Improving Regulatory Systems* and *Environmental Protection Agency*. Portions of a third, *Reinventing Environmental Management*, also bear on the EPA's responsibilities.

5. The office had a small staff of seven in fiscal 1993, and worked in the agency's administration and resources management division. It is assisted by an advisory committee consisting of community environmental groups, with some representation from academia, business, and government agencies. The office maintains an environmental justice hotline to receive calls from concerned citizens about justice issues in their communities.

6. The Clinton executive order is reprinted in Bullard (1994).
7. Other significant appointments include former Natural Resources Defense Council (NRDC) attorney Dan Reicher as deputy chief of staff and counselor at the DOE and NRDC attorney Mary Nichols as EPA assistant administrator for air and radiation. Former Senator Tim Wirth was named undersecretary of state for global affairs. Jessica T. Mathews, formerly vice president of the World Resources Institute, accepted a top position in Wirth's office. There are more than 400 positions in the environmental area eligible for presidential appointment. Many of these, however, are filled by career officials.
8. The League scores were reported in *E Magazine,* March–April, 1994. Greenpeace's comments were among other environmental group ratings covered in a news release from the Environment News Service, November 17, 1993, reporting on an article in *Mother Jones,* "Clinton's Green Card," in the November–December 1993 issue. In its November 6, 1993 issue, the National Journal's reporters reviewed key Clinton appointees, rating EPA head Carol Browner and Agriculture Secretary Mike Espy as deficient in management and innovation. Bruce Babbitt and Hazel O'Leary did better.
9. In May 1994, the House Science, Space, and Technology Committee held hearings on the NIE proposal. The published hearings will contain statements on the reasons for establishing the institute. A summary of the proposal and the findings of recent surveys of federal environmental research capabilities can be found in William K. Stevens, "Push for Environmental Institute," *New York Times,* March 8, 1994, 37.
10. The report (*Research to Protect, Restore, and Manage the Environment*) is discussed in NIE Network News, July 1993. An exception to the criticism of environmental research that is consistent with the NIE proposal is the 1991 Sustainable Biosphere Initiative of the Ecological Society of America. It offers an extensive agenda for research on global change, biodiversity, and sustainable ecological systems, and recognizes needs for public education and improved environmental decision making. See Lubchenco et al. (1991). Some further research and policy suggestions are found in a report on the ESA's joint project with the Society of Environmental Toxicology and Chemistry (SETAC) on sustainability-based environmental management (Barnthouse 1995).
11. For early reviews of UNCED's achievements and omissions, see Haas, Levy, and Parson (1992) and Soroos (1994). Fuller accounts are beginning to appear and more are likely over the next few years.
12. The potential of grassroots movements worldwide can be appreciated from a recent compilation. The volume *Environmental Profiles* offers detailed descriptions of some 7000 projects, programs, and campaigns on biodiversity, population planning, health, and sustainable development in 115 nations (Katz, Orrick and Honig 1992).
13. Many environmental groups have published lists of specific actions people may take to improve environmental quality. They usually number in the 100 plus range. See, for example, Zero Population Growth's "Making a Difference" (1989) and the *Utne Reader*'s "How the Environmental Crisis Can Improve Our Lives" (November–December 1989), 69–89. A fuller guide for citizens can be found in a Global Tomorrow Coalition handbook prepared for the twentieth Earth Day anniversary in 1990: *The Global Ecology Handbook: What You Can Do About the Environmental Crisis* (Boston: Beacon Press, 1990). It is one of the best collections available, with a comprehensive list of environmental organizations and concrete proposals for citizen action arranged by substantive problem area.

REFERENCES

Ackerman, Bruce A, and William T. Hassler. 1981. *Clean Coal/Dirty Air.* New Haven: Yale University Press.

Adler, Robert W. 1993. "Water Resources: Revitalizing the Clean Water Act." *Environment* 35 (November): 4–5, 40.

———. 1994. "The Clean Water Act: Has It Worked?" *EPA Journal* 20 (summer): 10–14.

Alliance to Save Energy, American Council for an Energy-Efficient Economy, Natural Resources Defense Council, and Union of Concerned Scientists. 1991. *America's Energy Choices: Investing in a Strong Economy and a Clean Environment.* Cambridge, Mass.: Union of Concerned Scientists.

Alper, Joseph. 1992. "War Over the Wetlands: Ecologists v. the White House." *Science* 257 (August 21): 1043–44.

Americans for the Environment. 1989. "The Rising Tide: Public Opinion, Policy and Politics." Louis Harris and Associates, conducted for Americans for the Environment, the Sierra Club, and the National Wildlife Federation, April 20.

Amy, Douglas J. 1987. *The Politics of Environmental Mediation.* New York: Columbia University Press.

Anderson, Charles W. 1979. "The Place of Principles in Policy Analysis." *American Political Science Review* 73 (September): 711–23.

Anderson, Curt. 1994. "EPA: Barely Half of Nation's Water Good for All Uses." Associated Press, April 20.

Anderson, James E. 1994. *Public Policymaking: An Introduction,* 2nd ed. Boston: Houghton Mifflin.

Anderson, Terry L., and Donald R. Leal. 1991. *Free Market Environmentalism.* Boulder: Westview Press.

Andrews, Richard N. L. 1994. "Risk-Based Decisionmaking." In Vig and Kraft, *Environmental Policy in the 1990s.*

Axelrod, Regina. 1984. "Energy Policy: Changing the Rules of the Game." In Vig and Kraft, *Environmental Policy in the 1980s.*

Balzhiser, Richard E., and John E. Bryson. 1994. "The Strategic Role of Electric Vehicles." *Forum for Applied Research and Public Policy* 9 (spring): 31–36.

Barnthouse, Larry, ed. 1995. *Sustainable Environmental Management.* Pensacola, Fla.: SETAC Foundation for Environmental Education, Proceedings of a Workshop at Pellston, Michigan, August 25–31, 1993.

Bartlett, Robert V. 1984. "The Budgetary Process and Environmental Policy." In Vig and Kraft, *Environmental Policy in the 1980s.*

———. 1986. "Ecological Rationality: Reason and Environmental Policy." *Environmental Ethics* 8 (fall): 221–40.

———, ed. 1989. *Policy Through Impact assessment: Institutionalized Analysis as a Policy Strategy.* New York: Greenwood Press.

———. 1990. "Comprehensive Environmental Decision Making: Can It Work?" In Vig and Kraft, *Environmental Policy in the 1990s.*

———. 1994. "Evaluating Environmental Policy Success and Failure." In Vig and Kraft, *Environmental Policy in the 1990s.*

Bartlett, Robert V., and Charles R. Malone, eds. 1993. "Science and the National Environmental Policy Act." *The Environmental Professional* 15 (1): 1–149.

Baskin, Yvonne. 1993. "Ecologists Dare to Ask: How Much Does Diversity Matter?" *Science* 264 (April 8): 202–203.

Baumgartner, Frank R., and Bryan D. Jones. 1993. *Agendas and Instability in American Politics.* Chicago: University of Chicago Press.

Bayard, Steven, and Jennifer Jinot. 1993. "Environmental Tobacco Smoke: Industry's Suit." *EPA Journal* 19 (October–December): 20.

Belzer, Richard B. 1991. "The Peril and Promise of Risk Assessment." *Regulation: Cato Review of Business and Government* 14 (fall): 40–49.

Benedick, Richard E. 1991. *Ozone Diplomacy.* Cambridge: Harvard University Press.

Berke, Richard. 1993. "Clinton Backs Off from Policy Shift on Federal Lands." *New York Times* (March 31): 1, A9.

Berry, Jeffrey M. 1977. *Lobbying for the People: The Political Behavior of Public Interest Groups.* Princeton, N.J.: Princeton University Press.

———. 1989. *The Interest Group Society,* 2nd ed. Glenview, Ill.: Scott, Foresman.

Berry, Jeffrey M., and Kent E. Portney. 1995. "Centralizing Regulatory Control and Interest Group Access: The Quayle Council on Competitiveness." In *Interest Group Politics,* 4th ed., edited by Allan J. Cigler and Burdett A. Loomis, eds., Washington.: CQ Press.

Berry, Jeffrey M., Kent E. Portney, and Ken Thomson. 1993. *The Rebirth of Urban Democracy.* Washington: Brookings Institution.

Better, Nancy Marx. 1992. "Green Teens." *New York Times Magazine* (March 8): 44, 66–68.

Bezdek, Roger H. 1993. "Environment and Economy: What's the Bottom Line." *Environment* 35 (September): 7–11, 25–32.

Bingham, Gail. 1986. *Resolving Environmental Disputes: A Decade of Experi-*ence. Washington.: Conservation Foundation.

Bowermaster, Jon. 1993. "Is Carol Browner in Over Her Head? *Audubon* (September–October): 59–63.

Bowman, Ann O'M., and Mark E. Tompkins. 1993. "Environmental Protection and Economic Development: Can States Have It Both Ways." Paper presented at the annual meeting of the American Political Science Association, Washington, D.C., September 2–5.

Bosso, Christopher J. 1987. *Pesticides and Politics: The Life Cycle of a Public Issue.* Pittsburgh: University of Pittsburgh Press.

———. 1991. "Adaptation and Change in the Environmental Movement." In *Interest Group Politics,* 3rd ed., edited by Allan J. Cigler and Burdett A. Loomis. Washington: CQ Press.

———. 1994. "After the Movement: Environmental Activism in the 1990s." In Vig and Kraft, *Environmental Policy in the 1990s.*

Brody, Annie. 1993. "Growing Up Green." *The Amicus Journal* (fall): 10–12.

Browner, Carol M. 1993. "Environmental Tobacco Smoke: EPA's Report." *EPA Journal* 19 (October–December): 18–19.

Bryant, Bunyan, and Paul Mohai, eds. 1992. *Race and the Incidence of Environmental Hazards: A Time for Discourse.* Boulder, Colo.: Westview Press.

Bryner, Gary C. 1987. *Bureaucratic Discretion: Law and Policy in Federal Regulatory Agencies.* New York; Pergamon Press.

———. 1993. *Blue Skies, Green Politics: The Clean Air Act of 1990.* Washington: CQ Press.

Bullard, Robert D. 1990. *Dumping in Dixie: Race, Class, and Environmental Quality.* Boulder, Colo.: Westview Press.

———, ed. 1993. *Confronting Environmental Racism: Voices from the Grassroots.* Boston: South End Press.

———. 1994. "Overcoming Racism in Environmental Decisionmaking." *Environment* 36 (May): 10–20, 39–44.

Cahn, Robert. 1993. "Report on Reports: Science and the National Parks." *Environment* 35 (March): 25–27.

Caldwell, Lynton Keith. 1970. *Environment: A Challenge for Modern Society.* Garden City, N.Y.: Natural History Press, Doubleday.

———. 1982. *Science and the National Environmental Policy Act: Redirecting Policy through Procedural Reform.* University, Ala.: University of Alabama Press.

———. 1990. *International Environmental Policy: Emergence and Dimensions,* 2nd ed. Durham: Duke University Press.

Calvert, Jerry W. 1989. "Party Politics and Environmental Policy." In Lester, *Environmental Politics and Policy.*

Camia, Catalina. 1993a. "Clinton's Forest Compromise Is Assailed from All Sides." *Congressional Quarterly Weekly Report,* July 3, 1726–67.

———. 1993b. "Poor, Minorities Want Voice in Environmental Choices." *Congressional Quarterly Weekly Report* (August 21): 2257–60.

———. 1994a. "Severity of Job Loss at Issue in Mining Law Overhaul." *Congressional Quarterly Weekly Report* (January 8): 18–20.

———. 1994b. "Complexities Face Congress in 'Superfund' Overhaul." *Congressional Quarterly Weekly Report* (February 5): 239–40.

———. 1994c. "Legislators Draw in the Reins on Environmental Rules." *Congressional Quarterly Weekly Report* (April 30): 1060–63.

Carnegie Commission. 1992. *Environmental Research and Development: Strengthening the Federal Infrastructure.* New York: Carnegie Commission on Science, Technology, and Government, December.

———. 1993a. *Science, Technology, and Government for a Changing World.* New York: Carnegie Commission on Science, Technology, and Government, April.

———. 1993b. *Risk and the Environment: Improving Regulatory Decision Making.* New York: Carnegie Commission on Science, Technology, and Government, June.

Carney, Eliza Newlin. 1994. "Lobbying Flood on Clean Water Bill." *National Journal* (March 19):664.

Catton, William R., Jr., and Riley E. Dunlap. 1980. "A New Ecological Paradigm for Post-Exuberant Sociology." *American Behavioral Scientist* 24: 15–47.

Chivian, Eric, Michael McCally, Howard Hu, and Andrew Haines. 1993. *Critical Condition: Human Health and the Environment.* Cambridge: MIT Press.

Clarke, Jeanne Nienaber, and Daniel McCool. 1985. *Staking Out the Terrain: Power Differentials Among Natural Resource Management Agencies.* Albany: State University of New York Press.

Clary, Bruce B., and Michael E. Kraft. 1989. "Environmental Assessment, Science, and Policy Failure: The Politics of Nuclear Waste Disposal." In Bartlett, *Policy Through Impact Assessment.*

Cohen, Richard E. 1992. *Washington at Work: Back Rooms and Clean Air.* New York: Macmillan.

Cohen, Steven. 1984. "Defusing the Toxic Time Bomb: Federal Hazardous Waste Programs." In Vig and Kraft, *Environmental Policy in the 1980s.*

Cohen, Steven, and Sheldon Kamieniecki. 1991. *Environmental Regulation Through Strategic Planning.* Boulder, Colo.: Westview.

Cole, Leonard A. 1993. *Element of Risk: The Politics of Radon.* Washington: AAAS Press.

Congressional Quarterly. 1983. *The Battle for Natural Resources.* Washington: Congressional Quarterly, Inc.

Cooper, Joseph, and William F. West. 1988. "Presidential Power and Republican Government: The Theory and Practice of OMB Review of Agency Rules." *Journal of Politics* 50 (November): 864–95.

Costanza, Robert, ed. 1991. *Ecological Economics: The Science and Management of Sustainability.* New York: Columbia University Press.

Costanza, Robert, Bryan G. Norton, and Benjamin D. Haskell, eds. 1992. *Ecosystem Health: New Goals for*

Ecosystem Management. Washington: Island Press.

Council on Environmental Quality. 1980. "Public Opinion on Environmental Issues: Results of a National Public Opinion Survey." Washington: CEQ.

———. 1993. *Environmental Quality: 23rd Annual Report of the Council on Environmental Quality.* Washington: CEQ, January.

Council on Environmental Quality and Department of State. 1980. *Global 2000 Report to the President,* Vols. I, II, and III. Washington: Government Printing Office.

Cowell, Alan. 1994. "U.N. Population Meeting Adopts Program of Action." *New York Times* (September 14):A2.

Cronon, William. 1983. *Changes in the Land: Indians, Colonists and the Ecology of New England.* New York: Hill and Wang.

Culhane, Paul J. 1981. *Public Lands Politics: Interest Group Influence on the Forest Service and the Bureau of Land Management.* Baltimore: Johns Hopkins University Press.

———. 1984. "Sagebrush Rebels in Office: Jim Watt's Land and Water Politics." In Vig and Kraft, *Environmental Policy in the 1980s.*

Cushman, John H., Jr. 1992. "Quayle, in Last Push for Landowners, Seeks to Relax Wetland Protections." *New York Times* (November 12): A8.

———. 1994a. "Administration Plans Revision to Ease Toxic Cleanup Criteria." *New York Times* (January 31):1, A7.

———. 1994b. "E.P.A. Critics Get Boost in Congress." *New York Times,* (May 7): 1, C9.

———. 1994c. "Forced, U.S. Sells Gold Land for Trifle." *New York Times* (May 17):A7.

———. 1994d. "Flood Panel Wants Emphasis on Evacuations, Not Dams." *New York Times* (May 26):1, A8.

———. 1994e. "Eagles to Fly Free of the Endangered List." *New York Times* (June 30):A8.

———. 1994f. "Department of Energy Pushes Competitive Bids." *New York Times* (July 7):C2.

———. 1994g. "Environmental Lobby Beats Tactical Retreat." *New York Times,* March 30, B7 (New York edition).

Daly, Herman E., and John B. Cobb, Jr. 1989. *For the Common Good: Redirecting the Economy Toward Community, the Environment, and a Sustainable Future.* Boston: Beacon Press.

Dana, Samuel Trask, and Sally K. Fairfax. 1980. *Forest and Range Policy: Its Development in the United States,* 2nd ed. New York: McGraw-Hill.

Darman, Richard. 1990. "Keeping America First: American Romanticism and the Global Economy." The Second Annual Albert H. Gordon Lecture, Harvard University, May 1.

Davies, J. Clarence III, and Barbara S. Davies. 1975. *The Politics of Pollution,* 2nd ed. Indianapolis: Bobbs-Merrill.

Davis, Phillip A. 1991. "Cry for Preservation, Recreation Changing Public Land Policy." *Congressional Quarterly Weekly Report* (August 3):2145–51.

De Saillan, Charles. 1993. "In Praise of Superfund." *Environment* 35 (October): 42–44.

DiIulio, John J., Jr., Gerald Garvey, and Donald F. Kettl. 1993. *Improving Government Performance: An Owner's Manual.* Washington: Brookings Institution.

Douglas, Mary, and Aaron Wildavsky. 1982. *Risk and Culture.* Berkeley: University of California Press.

Dower, Roger. 1990. "Hazardous Wastes." In Portney, *Public Policies for Environmental Protection.*

Downs, Anthony. 1972. "Up and Down with Ecology—The 'Issue-Attention Cycle.'" *The Public Interest* (Summer): 38–50.

Dryzek, John S. 1987. *Rational Ecology: Environment and Political Economy.* New York: Basil Blackwell.

Dryzek, John S., and James P. Lester. 1989. "Alternative Views of the Environmental Problematique." In Lester, ed., *Environmental Politics and Policy.*

Duffy, Robert J. 1994. "The Politics of Regulatory Distrust: The Quayle Council on Competitiveness and the Clean Air Act." Paper presented at the annual meeting of the Western Political Science Association, Albuquerque, March 10–12.

Dunlap, Riley E. 1987. "Polls, Pollution, and Politics Revisited: Public Opinion on the Environment in the Reagan Era." *Environment* 29 (July/August): 6–11, 32–37.

———. 1989. "Public Opinion and Environmental Policy." In Lester, ed., *Environmental Politics and Policy.*

———. 1991. "Public Opinion in the 1980s: Clear Consensus, Ambiguous Commitment." *Environment* 33 (October): 10–15, 32–37.

———. 1992. "Trends in Public Opinion Toward Environmental Issues: 1965–1990." In Dunlap and Mertig, eds., *American Environmentalism.*

Dunlap, Riley E., and Angela G. Mertig, eds. 1992. *American Environmentalism: The U.S. Environmental Movement, 1970–1990.* Philadelphia: Taylor and Francis.

Dunlap, Riley E., and Rik Scarce. 1991. "The Polls—Poll Trends: Environmental Problems and Protection." *Public Opinion Quarterly* 55 (winter): 650–72.

Dunlap, Riley E., George H. Gallup, Jr., and Alec M. Gallup. 1993. "Of Global Concern: Results of the Health of the Planet Survey." *Environment* 35, 9: 7–15, 33–40.

Dunlap, Riley E., Michael E. Kraft, and Eugene A. Rosa, eds. 1993. *Public Reactions to Nuclear Waste: Citizens' Views of Repository Siting.* Durham: Duke University Press.

Durant, Robert F. 1992. *The Administrative Presidency Revisited: Public Lands, the BLM, and the Reagan Revolution.* Albany: State University of New York Press.

Durning, Alan. 1992. *How Much Is Enough? The Consumer Society and the Future of the Earth.* New York: W. W. Norton.

Eads, George C., and Michael Fix. 1984. *Relief or Reform? Reagan's Regulatory Dilemma.* Washington: Urban Institute.

Eckersley, Robyn. 1992. *Environmentalism and Political Theory: Toward an Ecocentric Approach.* Albany: State University of New York Press.

Egan, Timothy. 1993a. "Sweeping Reversal of U.S. Land Policy Sought by Clinton." *New York Times* (February 24): 1, A9.

———. 1993b. "Wingtip 'Cowboys' in Last Stand to Hold on to Low Grazing Fees." *New York Times* (October 29): 1, A8.

———. 1994a. "Tight Logging Limit Set in Northwest." *New York Times* (February 24):A8.

———. 1994b. "Oregon, Foiling Forecasters, Thrives as It Protects Owls." *New York Times* (October 11):1, C20.

Ehrlich, Paul R., and Edward O. Wilson. 1991. "Biodiversity Studies: Science and Policy." *Science* 253 (August 16): 758–62.

Environmental Coalition. 1994. "How to Defend Our Environmental Laws: A Citizen Action Guide." New York: Environmental Coalition, July.

Erskine, Hazel. 1972. "The Polls: Pollution and Its Costs." *Public Opinion Quarterly* 36 (spring): 120–135.

Finkel, Adam M., and Dominic Golding, eds. 1994. *Worst Things First? The Debate over Risk-Based National Environmental Priorities.* Washington: Resources for the Future.

Fischer, Frank. 1990. *Technocracy and the Politics of Expertise.* Newbury Park, Calif.: Sage.

Flavin, Christopher, and Nicholas Lenssen. 1994. *Power Surge: Guide to the Coming Energy Revolution.* Washington: Worldwatch Institute.

Flavin, Christopher, and John E. Young. 1993. "Shaping the Next Industrial Revolution." In *State of the World 1993,* edited by Lester R. Brown, et al. New York: W. W. Norton.

Foran, Jeffery A., and Robert W. Adler, 1993. "Cleaner Water, But Not Clean Enough." *Issues in Science and Technology* 10(2): 33–39.

Foss, Philip O. 1960. *Politics and Grass.* Seattle: University of Washington Press.

Fox, Stephen. 1985. *The American Conservation Movement: John Muir and His Legacy.* Madison: University of Wisconsin Press. Originally published by Little, Brown, 1981.

Freeman, A. Myrick, III. 1990. "Water Pollution Policy." In Portney, *Public Policies for Environmental Protection.*

————. 1993. *The Measurement of Environmental and Resource Values: Theory and Methods.* Washington: Resources for the Future.

————. 1994. "Economics, Incentives, and Environmental Regulation." In Vig and Kraft, *Environmental Policy in the 1990s.*

Freemuth, John C. 1991. *Islands Under Siege: National Parks and the Politics of External Threats.* Lawrence: University of Kansas Press.

French, Hilary F. 1994. "Rebuilding the World Bank." In *State of the World 1994,* edited by Lester R. Brown, et al. New York: W. W. Norton.

Funke, Odelia. 1994. "National Security and the Environment." In Vig and Kraft, *Environmental Policy in the 1990s.*

Furlong, Scott R. 1992. "Interest Group Influence on Regulatory Policy." Paper presented at the 1992 annual meeting of the American Political Science Association, Chicago, September.

————. 1994. "The 1992 Regulatory Moratorium: Did It Make a Difference?" Paper presented to the annual meeting of the Midwest Political Science Association, Chicago, April 14–16.

Gardiner, David. 1994. "Does Environmental Policy Conflict with Economic Growth?" *Resources* 115 (spring): 20–21.

Goodwin, Craufurd D., ed. 1981. *Energy Policy in Perspective.* Washington: Brookings Institution.

Gore, Al. 1992. *Earth in the Balance: Ecology and the Human Spirit.* Boston: Houghton Mifflin.

————. 1994. "Environmental Technologies for a Sustainable Future." *EPA Journal* 20 (fall): 6–8.

Gormley, William T., Jr. 1989. *Taming the Bureaucracy: Muscles, Prayers, and Other Strategies.* Princeton: Princeton University Press.

Gottlieb, Robert. 1993. *Forcing the Spring: The Transformation of the American Environmental Movement.* Washington: Island Press.

Gramp, Kathleen M. 1994. "Federal Funding for Environmental Research." In *Science and Technology Policy Yearbook 1993,* edited by Albert H. Teich, Stephen D. Nelson, and Celia McEnaney. Washington: American Association for the Advancement of Science.

Greve, Michael S., and Fred L. Smith, Jr. eds. 1992. *Environmental Politics: Public Costs, Private Rewards.* New York: Praeger.

Gurr, Ted Robert. 1985. "On the Political Consequences of Scarcity and Economic Decline." *International Studies Quarterly* 29: 51–75.

Haas, Peter M., Marc A. Levy, and Edward A. Parson. 1992. "Appraising the Earth Summit: How Should We Judge UNCED's Success?" *Environment* 34 (October): 6–11, 26–33.

Hadden, Susan G. 1986. *Read the Label: Reducing Risk by Providing Information.* Boulder, Colo.: Westview Press.

————. 1989. *A Citizen's Right to Know: Risk Communication and Public Policy.* Boulder, Colo.: Westview Press.

————. 1991. "Public Perception of Hazardous Waste." *Risk Analysis* 11, 1: 47–57.

Hager, Carol J. 1994. *Technological Democracy: Bureaucracy and Citizenry in the Germany Energy Debate.* Ann Arbor: University of Michigan Press.

Hall, Bob, and Mary Lee Kerr. 1991. *1991–92 Green Index: A State-By-State Guide to the Nation's Environmental Health.* Washington: Island Press.

Halley, Alexis A. 1994. "Hazardous Waste Disposal: The Double-Edged Sword of the RCRA Land-Ban Hammers." In *Who Makes Public Policy? The Struggle for Control Between Congress and the Exec-*

utive, edited by Robert S. Gilmour and Alexis A. Halley. Chatham, N.J.: Chatham House.

Hamilton, Michael S., ed. 1990. *Regulatory Federalism, Natural Resources, and Environmental Management.* Washington: American Society for Public Administration.

Hardin, Garrett. 1968. "The Tragedy of the Commons." *Science* 162 (December 13): 1243–48.

Harris, Richard A., and Sidney M. Milkis. 1989. *The Politics of Regulatory Change: A Tale of Two Agencies.* New York: Oxford University Press.

Harrison, Kathryn. 1995. "Is Cooperation the Answer? Canadian Environmental Enforcement in Comparative Context." *Journal of Policy Analysis and Management* 14 (spring): 221–44.

Hawken, Paul. 1994. *The Ecology of Commerce: A Declaration of Sustainability.* New York: HarperBusiness Publishers.

Hays, Samuel P. 1959. *Conservation and the Gospel of Efficiency.* Cambridge, England: Cambridge University Press.

———. 1987. *Beauty, Health, and Permanence: Environmental Politics in the United States, 1955–1985.* New York: Cambridge University Press.

Heilbroner, Robert L. 1991. *An Inquiry into the Human Prospect: Looked at Again for the 1990s.* New York: W. W. Norton.

Hempel, Lamont C. 1993. "Greenhouse Warming: The Changing Climate in Science and Politics." *Political Research Quarterly* 46, 1 (March): 213–39.

Henning, Daniel H., and William R. Mangun. 1989. *Managing the Environmental Crisis: Incorporating Competing Values in Natural Resource Administration.* Durham: Duke University Press.

Hershey, Robert D., Jr. 1992. "Regulations March on, Despite a Moratorium." *New York Times* (September 21): C1, C3.

Hill, Stuart. 1992. *Democratic Values and Technological Choices.* Palo Alto: Stanford University Press.

Hilts, Philip J. 1994. "Millions Live With Particles That Exceed U.S. Air Code." *New York Times* (April 30): 10.

Hird, John A. 1994. *Superfund: The Political Economy of Risk.* Baltimore: Johns Hopkins University Press.

Hirschhorn, Joel S., and Kirsten U. Oldenburg. 1991. *Prosperity Without Pollution: The Prevention Strategy for Industry and Consumers.* New York: Van Nostrand Reinhold.

Hoffman, David. 1990. "George Bush Has His Own Environmental Problems." *Washington Post National Weekly Edition,* April 30–May 6, 12.

Hogan, William H. 1984. "Energy Policy." In Portney, *Natural Resources and the Environment.*

Holdren, John P. 1991. "Energy in Transition." In *Energy for Planet Earth.* New York: W. H. Freeman.

Holusha, John. 1988. "Bush Pledges Aid for Environment." *New York Times* (September 1): 9.

———. 1994. "Bracing for the Worst in Chemicals." *New York Times* (June 4): 17, 27.

Horowitz, David. 1990. "Making the Green One Red." *National Review* (March 19): 39–40.

Howe, H. F. 1993. "The National Institute for the Environment: Comparison with Other Proposals. *The Environmental Professional* 15: 428–35.

Hunter, Susan, and Richard W. Waterman. 1992. "Determining an Agency's Regulatory Style: How Does the EPA Water Office Enforce the Law." *Western Political Quarterly* 45: 403–17.

Idelson, Holly. 1992a. "After Two-Year Odyssey, Energy Strategy Clears." *Congressional Quarterly Weekly Report* (October 10): 1992, 3141–46.

———. 1992b. "National Energy Strategy Provisions." *Congressional Quarterly Weekly Report,* November 28, 3722–3730.

Inglehart, Ronald. 1990. *Culture Shift in Advanced Industrial Society.* Princeton: Princeton University Press.

Ingram, Helen M., and Dean E. Mann. 1983. "Environmental Protection Policy." In *Encyclopedia of Policy Studies,* edited by Stuart S. Nagel. New York: Marcel Dekker.

Ingram, Helen, and Steven Rathgeb Smith, eds. 1993. *Public Policy for Democracy*. Washington: Brookings Institution.

Intergovernmental Panel on Climate Change. 1990. *Climate Change: The IPCC Scientific Assessment*. Geneva: World Meteorological Organization and United Nations Environment Programme.

Jacobson, Louis. 1993. "Green Giants." *National Journal* (May 8): 1113–16.

John, DeWitt. 1994. *Civic Environmentalism: Alternatives to Regulation in States and Communities*. Washington: CQ Press.

Johnson, Kirk. 1993. "Reconciling Rural Communities and Resource Conservation." *Environment* (November): 16–20, 27–33.

Johnston, David. 1993. "White House Urges Broad Protections for U.S. Wetlands." *New York Times* (August 25): 1, A9.

Jones, Charles O. 1975. *Clean Air: The Policies and Politics of Pollution Control*. Pittsburgh: University of Pittsburgh Press.

———. 1984. *An Introduction to the Study of Public Policy,* 3rd ed. Monterey, Calif.: Brooks/Cole.

Jordan, Andrew. 1994a. "Financing the UNCED Agenda: The Controversy over Additionality." *Environment* 36 (April): 16–20, 26–34.

———. 1994b. "Paying the Incremental Costs of Global Environmental Protection: The Evolving Role of GEF." *Environment* 36 (July–August): 12–20, 31–36.

Kamieniecki, Sheldon. 1995. "Political Parties and Environmental Policy." In Lester, *Environmental Politics and Policy*.

———, ed. 1993. *Environmental Politics in the International Arena: Movements, Parties, Organizations, and Policy*. Albany: State University of New York Press.

Katz, Linda Sobel, Sarah Orrick, and Robert Honig. 1992. *Environmental Profiles: A Global Guide to Projects and People*. Hamden, Conn.: Garland Publishing.

Kenski, Henry C., and Margaret Corgan Kenski. 1984. "Congress Against the President: The Struggle Over the Environment." In Vig and Kraft, *Environmental Policy in the 1980s*.

Kenworthy, Tom. 1992a. "Federal Projects Are Not Endangered." *Washington Post National Weekly Edition* (March 9–15): 37.

———. 1992b. "It's No Day at the Beach for the National Park Service." *Washington Post National Weekly Edition* (April 13–19): 34.

Kerr, J. B., and C. T. McElroy. 1993. "Evidence of Large Upward Trends of Ultraviolet-B Radiation Linked to Ozone Depletion." *Science* 262 (November 12): 1032–34.

Kerwin, Cornelius M., and Scott R. Furlong. 1992. "Time and Rulemaking: An Empirical Test of Theory." *Journal of Public Administration* 2 (April): 113–38.

Kettl, Donald F. 1993. *Sharing Power: Public Governance and Private Markets*. Washington: Brookings Institution.

Keyfitz, Nathan. 1990. "The Growing Human Population." In *Managing Planet Earth*. New York: W. H. Freeman.

Kingdon, John W. 1984. *Agendas, Alternatives, and Public Policies*. Boston: Little, Brown.

Knaap, Gerrit, T., John Kim, and John Fitipaldi, eds. 1996. *Environmental Program Evaluation: A Primer*. Champaign: University of Illinois Press, in press.

Knopman, Debra S, and Richard A. Smith. 1993. "Twenty Years of the Clean Water Act." *Environment* 35 (January–February): 17–20, 34–41.

Koplow, Douglas N. 1993. *Federal Energy Subsidies: Energy, Environmental, and Fiscal Impacts*. Lexington, Mass.: Alliance to Save Energy, April.

Kraft, Michael E. 1984. "A New Environmental Policy Agenda: The 1980 Presidential Campaign and Its Aftermath." In Vig and Kraft, *Environmental Policy in the 1980s*.

———. 1992. "Ecology and Political Theory: Broadening the Scope of Environmental Politics." *Policy Studies Journal* 20 (4): 712–18.

———. 1993. "Air Pollution in the West: Testing the Limits of Public Support with Southern California's Clean Air Policy." In *Environmental Politics and Policy in the West,* edited by Zachary Smith, 137–57. Dubuque, Iowa: Kendall-Hunt Publishers.

———. 1994a. "Environmental Gridlock: Searching for Consensus in Congress." In Vig and Kraft, *Environmental Policy in the 1990s.*

———. 1994b. "Population Policy." In *Encyclopedia of Policy Studies,* 2nd ed., edited by Stuart S. Nagel. New York: Marcel Dekker.

———. 1994c. "Searching for Policy Success: Reinventing the Politics of Site Remediation." *The Environmental Professional* 16 (September): 245–53.

———. 1995a. "Congress and Environmental Policy." In Lester, *Environmental Politics and Policy.*

———. 1995b. "Democratic Dialogue and Acceptable Risks: The Politics of High-Level Nuclear Waste Disposal in the United States." In *Critical Cases in Hazardous Waste Facility Siting: Seeds of Success and Roots of Failure,* edited by Don Munton. Washington: Georgetown University Press, in press.

Kraft, Michael E., and Bruce B. Clary. 1991. "Citizen Participation and the NIMBY Syndrome: Public Response to Radioactive Waste Disposal." *Western Political Quarterly* 44 (June): 299–328.

Kraft, Michael E., and Norman J. Vig. 1990. "Presidential Styles and Substance: Environmental Policy from Reagan to Bush." Paper presented at the annual meeting of the American Political Science Association, San Francisco, August 30–September 2.

Kraft, Michael E., Bruce B. Clary, and Richard J. Tobin. 1988. "The Impact of New Federalism on State Environmental Policy: The Great Lakes States." In *The Midwest Response to the New Federalism,* edited by Peter Eisinger and William Gormley. Madison: University of Wisconsin Press.

Krimsky, Sheldon, and Alonzo Plough. 1988. *Environmental Hazards: Communicating Risks as a Social Process.* Dover, Mass.: Auburn House.

Kriz, Margaret. 1993. "Quick Draw." *National Journal* (November 13): 2711–16.

———. 1994a. "Super Fight." *National Journal* (January 29): 224–29.

———. 1994b. "Cleaner Than Clean?" *National Journal* (April 23): 946–49.

———. 1994c. "What's the Point of Finger-Pointing." *National Journal* (May 7): 1097.

———. 1994d. "How the Twain Met." *National Journal* (June 4): 1291–95.

———. 1994e. "Drill or Die." *National Journal* (January 8): 73–76.

———. 1994f. "The Greening of Environmental Regulation." *National Journal* (June 18): 1464–67.

Kusler, Jon A., William J. Mitsch, and Joseph S. Larson. 1994. "Wetlands." *Scientific American* 220 (January): 64–70.

Lacey, Michael J., ed. 1989. *Government and Environmental Politics: Essays on Historical Developments Since World War II.* Baltimore: Johns Hopkins University Press.

Lancelot, Jill, and Ralph DeGennaro. 1995. "'Green Scissors' Snip $33 Billion." *New York Times* (January 31): A11.

Landy, Marc K, Marc J. Roberts, and Stephen R. Thomas. 1990. *The Environmental Protection Agency: Asking the Wrong Questions.* New York: Oxford University Press.

Lave, Lester B. 1981. *The Strategy of Social Regulation: Decision Frameworks for Policy.* Washington: Brookings Institution.

League of Conservation Voters. 1994. "National Environmental Scorecard." Washington: League of Conservation Voters, February.

Leary, Warren W. 1994. "Studies Raise Doubts About Need to Lower Home Radon Levels." *New York Times* (September 6): B7.

Lee, Gary. 1994. "A Potential Killer's Modus Operandi." *Washington Post National Weekly Edition* (June 20–26): 38.

Leiss, William. 1976. *The Limits to Satisfaction: An Essay on the Problem of Needs and Commodities.* Toronto: University of Toronto Press.

Lenssen, Nicholas. 1993. "Providing Energy in Developing Countries." In *State of the World 1993,* edited by Lester R. Brown et al. New York: W. W. Norton.

Leopold, Aldo. 1949. *A Sand County Almanac.* New York: Oxford University Press; exp. ed., with additional essays from *Round River,* N.Y.: Ballantine Books, 1970.

Leshy, John D. 1984. "Natural Resource Policy." In Portney, *Natural Resources and the Environment.*

Lester, James P. 1994. "A New Federalism? Environmental Policy in the States." In Vig and Kraft, *Environmental Policy in the 1990s.*

———, ed. 1989. *Environmental Politics and Policy: Theories and Evidence.* Durham: Duke University Press.

———, ed. 1995. *Environmental Politics and Policy: Theories and Evidence,* 2nd ed. Durham: Duke University Press.

Lewis, Martin W. 1994. *Green Delusions: An Environmentalist Critique of Radical Environmentalism.* Durham: Duke University Press.

Lewis, Paul. 1994. "U.N. Panel Finds Action on Environment Lagging." *New York Times* (May 29): 6.

Lindblom, Charles E. 1959. "The Science of 'Muddling Through.'" *Public Administration Review* 19: 79–88.

Lindblom, Charles E., and Edward J. Woodhouse. 1993. *The Policy-Making Process,* 3rd ed. Englewood Cliffs, N.J.: Prentice-Hall.

Lindzen, Richard S. 1992. "Global Warming: The Origin and Nature of the Alleged Scientific Consensus." *Regulation: Cato Review of Business and Government* 15 (spring): 87–98.

Lippman, Thomas W. 1993. "Generating Electricity by Not Using It." *Washington Post National Weekly Edition* (March 29–April 4):34.

Lowi, Theodore J. 1979. *The End of Liberalism,* 2nd ed. New York: W. W. Norton.

Lowry, William R. 1992. *The Dimensions of Federalism: State Governments and Pollution Control Policies.* Durham: Duke University Press.

———. 1994. *The Capacity for Wonder: Preserving National Parks.* Washington: Brookings Institutions.

Lubchenco, Jane, et al. 1991. "The Sustainable Biosphere Initiative: An Ecological Research Agenda." *Ecology* 72 (April): 371–412.

Lutz, Ernst, 1993. *Toward Improved Accounting for the Environment.* Washington: The World Bank.

Lutz, Wolfgang. 1994. "The Future of World Population." *Population Bulletin* 49 (June): 1–47.

MacNeill, Jim, Pieter Winsemius, and Taizo Yakushiji. 1991. *Beyond Interdependence: The Meshing of the World's Economy and the Earth's Ecology.* New York: Oxford University Press.

Mann, Charles C., and Mark L. Plummer. 1993. "The High Cost of Biodiversity." *Science* 260 (June 25): 1868–71.

Mann, Dean E. 1986. "Democratic Politics and Environmental Policy." In *Controversy in Environmental Policy,* edited by Sheldon Kamieniecki, Robert O'Brien, and Michael Clarke. Albany: State University of New York Press.

Marcus, Alfred A. 1992. *Controversial Issues in Energy Policy.* Newbury Park, Calif.: Sage.

Mathews, Jessica Tuchman. 1991a. "The Implications for U.S. Policy." In Mathews, *Preserving the Global Environment.*

———. ed. 1991b. *Preserving the Global Environment: The Challenge of Shared Leadership.* New York: W. W. Norton.

Mazmanian, Daniel. 1992. "Toward a New Energy Paradigm." In *California Policy Choices,* vol. 8, edited by John J. Kirlin. Los Angeles: University of Southern California, School of Public Administration.

Mazmanian, Daniel, and David Morell. 1992. *Beyond Superfailure: America's Toxics Policy for the 1990s.* Boulder, Colo.: Westview Press.

Mazmanian, Daniel A., and Jeanne Nienaber. 1979. *Can Organizations Change? Environmental Protection, Citizen Participation, and the Corps of Engineers.* Washington: Brookings Institution.

Mazmanian, Daniel A., and Paul A. Sabatier. 1983. *Implementation and Public Policy.* Glenview, Ill.: Scott, Foresman.

Mazur, Laurie Ann. 1994. *Beyond the Numbers: A Reader on Population, Consumption, and the Environment.* Washington: Island Press.

McConnell, Grant. 1966. *Private Power and American Democracy.* New York: Alfred Knopf.

McCool, Daniel. 1990. "Subgovernments as Determinants of Political Viability." *Political Science Quarterly* 105 (summer): 269–93.

McCormick, John. 1989. *Reclaiming Paradise: The Global Environmental Movement.* Bloomington: Indiana University Press.

McInnis, Doug. 1993. "Higher Grazing Fees Have Ranchers Running Scared." *New York Times* (September 12): F5.

Meadows, Donella H., Dennis L. Meadows, and Jørgen Randers. 1992. *Beyond the Limits: Confronting Global Collapse, Envisioning a Sustainable Future.* Post Mills, Vt.: Chelsea Green.

Melnick, R. Shep. 1983. *Regulation and the Courts: The Case of the Clean Air Act.* Washington: Brookings Institution.

Meyer, Stephen M. 1993. "Environmentalism and Economic Prosperity." Unpublished manuscript. Massachusetts Institute of Technology, Department of Political Science.

Michaelis, Laura. 1993. "Calls for Pesticide Reform Linked to New Report." *Congressional Quarterly Weekly Report* (July 3): 1730.

Milbrath, Lester W. 1984. *Environmentalists: Vanguard for a New Society.* Albany: State University of New York Press.

———. 1989. *Envisioning a Sustainable Society: Learning Our Way Out.* Albany: State University of New York Press.

Miller, Kenton R., Walter V. Reid, and Charles V. Barber. 1991. "Deforestation and Species Loss: Responding to the Crisis." In Mathews, *Preserving the Global Environment.*

Misch, Ann. 1994. "Assessing Environmental Health Risks." In *State of the World 1994,* edited by Lester R. Brown, et al. New York: W. W. Norton.

Mitchell, Robert Cameron. 1984. "Public Opinion and Environmental Politics in the 1970s and 1980s." In Vig and Kraft, *Environmental Policy in the 1980s.*

———. 1989. "From Conservation to Environmental Movement: The Development of the Modern Environmental Lobbies." In Lacey, *Government and Environmental Politics.*

———. 1990. "Public Opinion and the Green Lobby: Poised for the 1990s?" In Vig and Kraft, *Environmental Policy in the 1990s.*

Mitchell, Robert Cameron, Angela G. Mertig, and Riley E. Dunlap. 1992. "Twenty Years of Environmental Mobilization: Trends Among National Environmental Organizations." In Dunlap and Mertig, *American Environmentalism.*

Morone, Joseph G., and Edward J. Woodhouse. 1986. *Averting Catastrophe: Strategies for Regulating Risky Technologies.* Berkeley: University of California Press.

Nash, Roderick Frazier. 1989. *The Rights of Nature: A History of Environmental Ethics.* Madison: University of Wisconsin Press.

———. 1990. *American Environmentalism: Readings in Conservation History,* 3rd. ed. New York: McGraw-Hill.

National Academy of Public Administration. 1987. *Presidential Management of Rulemaking in Regulatory Agencies.* Washington: National Academy of Public Administration, January.

National Academy of Sciences. 1991. *Policy Implications of Greenhouse Warming.* Washington: National Academy Press.

National Acid Precipitation Assessment Program. 1990. *Background on Acidic Deposition and the National Acid Precipitation Assessment Program and Assessment Highlights.* Washington: NAPAP, September.

National Commission on the Environment. 1993. *Choosing a Sustainable Future: The Report of the National Commission on the Environment.* Washington: Island Press.

National Journal. 1990. "Opinion Outlook: Views on the American Scene." *National Journal* (April 28): 1052.

National Performance Review. 1993a. *From Red Tape to Results: Creating a Government That Works Better and Costs Less.* Washington: Office of the Vice President, September.

———. 1993b. *Accompanying Report of the National Performance Review,* multiple reports. Washington: Office of the Vice President, September (released in 1994).

National Research Council. 1983. *Risk Assessment in the Federal Government: Managing the Process.* Washington: National Academy Press.

———. 1990. *Forestry Research: A Mandate for Change.* Washington: National Academy Press.

New York Times. 1994a. "Environmental President? Not Yet." *New York Times* (February 13): E14.

———. 1994b. "U.S. Issues New Rules on Protected Species." *New York Times* (June 15): 7.

Norse, Elliott A. 1990. "What Good Are Ancient Forests?" *The Amicus Journal* (winter): 42–45.

O'Callaghan, Kate. 1992. "Whose Agenda for America? *Audubon* (September–October): 80–91.

Olson, Mancur. 1971. *The Logic of Collective Action.* Cambridge: Harvard University Press.

Ophuls, William, and A. Stephen Boyan Jr. 1992. *Ecology and the Politics of Scarcity Revisited.* New York: W. H. Freeman.

Orr, David W. 1992. *Ecological Literacy: Education and the Transition to a Postmodern World.* Albany: State University of New York Press.

Osborne, David, and Ted Gaebler. 1992. *Reinventing Government: How the Entrepreneurial Spirit Is Transforming the Public Sector.* Reading, Mass.: Addison-Wesley.

O'Toole, Randall. 1988. *Reforming the Forest Service.* Washington: Island Press.

Paehlke, Robert C. 1989. *Environmentalism and the Future of Progressive Politics.* New Haven: Yale University Press.

———. 1994. "Environmental Values and Public Policy." In Vig and Kraft, *Environmental Policy in the 1990s.*

Passell, Peter. 1993. "Polls May Help Government Decide the Worth of Nature." *New York Times* (September 6): 1, 20.

Peskin, Henry M., Paul R. Portney, and Allen V. Kneese, eds. 1981. *Environmental Regulation and the U.S. Economy.* Baltimore: Johns Hopkins University Press.

Poole, William. 1992. "Neither Wise Nor Well." *Sierra* (November–December): 59–61, 88–93.

Population Reference Bureau. 1994. "1994 World Population Data Sheet." Washington: Population Reference Bureau, Inc., April.

Porter, Gareth, and Janet Welsh Brown. 1991. *Global Environmental Politics.* Boulder Colo.: Westview Press.

Portney, Kent. E. 1992. *Controversial Issues in Environmental Policy: Science vs. Economics vs. Politics.* Newbury Park, Calif.: Sage.

Portney, Paul R. 1990b. "Air Pollution Policy." In Portney, *Public Policies for Environmental Protection.*

———. 1994. "Does Environmental Policy Conflict with Economic Growth?" *Resources,* No. 115 (spring): 21–23.

———. 1995. "Beware of the Killer Clauses Inside the GOP's 'Contract.'" *Washington Post National Weekly Edition* (January 23–29): 21.

————. ed. 1984. *Natural Resources and the Environment: The Reagan Approach.* Washington: Urban Institute Press.

————. ed. 1990a. *Public Policies for Environmental Protection.* Washington: Resources for the Future.

Portney, Paul R., and Katherine N. Probst. 1994. "Cleaning Up Superfund." *Resources,* No. 114 (winter): 2–5.

Postel, Sandra. 1988. "Controlling Toxic Chemicals." In *State of the World 1988,* edited by Lester R. Brown et al. New York: W. W. Norton.

————. 1994. "Carrying Capacity: Earth's Bottom Line." In *State of the World 1994,* edited by Lester R. Brown et al. New York: W. W. Norton.

Rabe, Barry G. 1986. *Fragmentation and Integration in State Environmental Management.* Washington: Conservation Foundation.

————. 1990. "Legislative Incapacity: The Congressional Role in Environmental Policy-making and the Case of Superfund." *Journal of Health Politics, Policy and Law* 15: 571–89.

————. 1994. *Beyond the NIMBY Syndrome: Hazardous Waste Facility Siting in Canada and the United States.* Washington: Brookings Institution.

Regan, Mary Beth. 1993a. "Uncle Sam Goes on an Eco-Trip." *Business Week* (June 28): 76.

————. 1993b. "The Sun Shines Brighter on Alternative Energy." *Business Week* (November 8): 94–95.

Reinhold, Robert. 1993a. "Hard Times Dilute Enthusiasm for Clean-Air Laws." *New York Times* (November 26): 1, A12.

————. 1993b. "U.S. Moves to Divert Fresh Water to Save Big Estuary in California." *New York Times* (December 16): 1, A12.

Renner, Michael. 1991. "Jobs in a Sustainable Economy." Washington: Worldwatch Institute, Worldwatch Paper 104, September.

Ringquist, Evan J. 1993. *Environmental Protection at the State Level: Politics and Progress in Controlling Pollution.* Armonk, N.Y. M. E. Sharpe.

Ripley, Randall B., and Grace A. Franklin. 1991. *Congress, the Bureaucracy, and Public Policy,* 5th ed. Pacific Grove, Calif.: Brooks/Cole.

Rocky Mountain Institute. 1992. "Efficient Car Revolution Accelerates." *Rocky Mountain Institute Newsletter* 8 (spring): 5.

Roper Organization. 1990. *The Environment: Public Attitudes and Individual Behavior.* Roper Organization, Inc. Commissioned by S. C. Johnson & Sons, July.

Rosenbaum, Walter A. 1987. *Energy, Politics, and Public Policy,* 2nd ed. Washington: CQ Press.

————. 1991. *Environmental Politics and Policy,* 2nd ed. Washington: CQ Press.

————. 1994. "The Clenched Fist and the Open Hand: Into the 1990s at EPA." In Vig and Kraft, *Environmental Policy in the 1990s.*

Ruckelshaus, William D. 1990. "Toward a Sustainable World." In *Managing Planet Earth.* New York: W. H. Freeman.

Rushefsky, Mark. 1986. *Making Cancer Policy.* Albany: State University of New York Press.

Russell, Clifford S. 1990. "Monitoring and Enforcement." In Portney, *Public Policies for Environmental Protection.*

Russell, Dick. 1994. "From Bureaucracy to Brave New World?" *The Amicus Journal* (summer): 27–31.

Russell, Milton, E. William Colglazier, and Bruce E. Tonn. 1992. "The U.S. Hazardous Waste Legacy." *Environment* 34: 12–15, 34–39.

Sabatier, Paul A., and Hank C. Jenkins-Smith. 1993. *Policy Change and Learning: An Advocacy Coalition Approach.* Boulder, Colo.: Westview.

Sagoff, Mark. 1988. *The Economy of the Earth.* Cambridge, England: Cambridge University Press.

————. 1993. "Environmental Economics: An Epitaph." *Resources* 111 (spring): 2–7.

Samet, Jonathan M., and John D. Spengler, eds. 1991. *Indoor Air Pollution: A Health Perspective.* Baltimore: Johns Hopkins University Press.

Schattschneider, E. E. 1960. *The Semi-Sovereign People: A Realist's View of Democracy in America.* New York: Holt, Rinehart and Winston.

Schlozman, Kay Lehman., and John T. Tierney. 1986. *Organized Interests and American Democracy.* New York: Harper and Row.

Schmidheiny, Stephan. 1992. *Changing Course: A Global Business Perspective on Development and the Environment.* Cambridge: MIT Press.

Schneider, Anne L., and Helen Ingram. 1990. "Policy Design: Elements, Premises, and Strategies." In *Policy Theory and Policy Evaluation: Concepts, Knowledge, Causes, and Norms,* edited by Stuart S. Nagel. Westport, Conn.: Greenwood Press.

Schneider, Keith. 1991. "Ozone Depletion Harming Sea Life." *New York Times* (November 16): 6.

———. 1992a. "Administration Tries to Limit Rule Used to Halt Logging of National Forests." *New York Times* (April 28): A7.

———. 1992b. "Industries Gaining Broad Flexibility on Air Pollution." *New York Times* (June 26): 1, A10.

———. 1993a. "Second Chance on Environment." *New York Times* (March 26): A11.

———. 1993b. "Administration's Pesticide Plan Puts Safety First." *New York Times* (September 21): 1, A12.

———. 1993c. "Pesticide Plan Could Uproot U.S. Farming." *New York Times* (October 10): 6.

———. 1993d. "Unbending Regulations Incite Move to Alter Pollution Laws." *New York Times* (November 29): 1, A11.

———. 1993e. "How a Rebellion Over Environmental Rules Grew from a Patch of Weeds." *New York Times* (March 24): C19.

———. 1994a. "Incinerators' Users Say Ruling Will Be Costly." *New York Times* (May 3): A12.

———. 1994b. "Progress, Not Victory, on Great Lakes Pollution." *New York Times* (May 7): 1, 9.

———. 1994c. "E.P.A. Moves to Reduce Health Risks From Dioxin." *New York Times* (September 14): A8.

———. 1994d. "Exxon Is Ordered to Pay $5 Billion for Alaska Spill." *New York Times* (September 17): 1, 9.

———. 1995. "Fighting to Keep U.S. Rules from Devaluing Land." *New York Times* (January 9): 1, A8.

Schneider, Stephen H. 1990. "The Changing Climate." In *Managing Planet Earth.* New York: W. H. Freeman.

Seelye, Katharine Q. 1994. "Voters Disgusted with Politicians As Election Nears." *New York Times* (November 3): 1, A28.

Shabecoff, Philip. 1989. "U.S. Only Narrowly Avoided 17 Bhopal-Like Disasters, Study Says." *New York Times* (April 30): 16.

———. 1993. *A Fierce Green Fire: The American Environmental Movement.* New York: Hill and Wang.

Shanley, Robert A. 1992. *Presidential Influence and Environmental Policy.* Westport, Conn.: Greenwood Press.

Shapiro, Michael. 1990. "Toxic Substances Policy." In Portney, *Public Policies for Environmental Protection.*

Shrader-Frechette, K. S. 1991. *Risk and Rationality: Philosophical Foundations for Populist Reforms.* Berkeley: University of California Press.

———. 1993. *Burying Uncertainty: Risk and the Case Against Geological Disposal of Nuclear Waste.* Berkeley: University of California Press.

Simon, Julian L., and Herman Kahn, eds. 1984. *The Resourceful Earth: A Response to "Global 2000."* New York: Basil Blackwell.

Slovic, Paul. 1987. "Perception of Risk." *Science* 236: 280–85.

———. 1993. "Perceived Risk, Trust, and Democracy." *Risk Analysis* 13: 675–82.

Smothers, Ronald. 1994. "Group Urges Tough Limits on Logging." *New York Times* (May 27): A9.

Soroos, Marvin. 1994. "From Stockholm to Río: The Evolution of Global Environmental Governance." In Vig and Kraft, *Environmental Policy in the 1990s.*

South Coast Air Quality Management District and the Southern California Association of Governments. 1989. *Air Quality Management Plan: South Coast Air Basin*. El Monte, Calif.: South Coast Air Quality Management District, March.

Starke, Linda. 1990. *Signs of Hope: Working Towards Our Common Future*. New York: Oxford University Press.

Stavins, Robert N. 1991. *Project 88—Round II, Incentives for Action: Designing Market–Based Environmental Strategies*. Washington: A public policy study sponsored by Senator Timothy E. Wirth and Senator John Heinz, May.

Stevens, William K. 1991. "Panel Urges Big Wetlands Restoration Project." *New York Times* (December 12): A16.

———. 1993. "Scientists Confront Renewed Backlash on Global Warming. *New York Times* (September 14): B5–B6.

———. 1994a. "Money Grow on Trees? No, But Study Finds Next Best Thing." *New York Times* (April 12): B12.

———. 1994b. "Emissions Must Be Cut to Avert Shift in Climate, Panel Says." *New York Times* (September 20): B9.

Stone, Deborah A. 1988. *Policy Paradox and Political Reason*. Glenview, Ill: Scott, Foresman.

Stone, Richard. 1993. "Babbitt Shakes Up Science at Interior." *Science* 261 (August 20): 976–78.

———. 1994a. "Study Implicates Second-Hand Smoke." *Science* 264 (April 1): 30.

———. 1994b. "Browner to Beef Up Outside Research." *Science* 265 (July 29): 599–600.

———. 1994c. "California Report Sets Standard for Comparing Risks." *Science* 266 (October 14): 214.

Stroup, Richard L., and John A. Baden. 1983. *Natural Resources: Bureaucratic Myths and Environmental Management*. San Francisco: Pacific Institute for Public Policy Research.

Susskind, Lawrence. 1987. *Breaking the Impasse: Consensual Approaches to Resolving Environmental Disputes*. New York: Basic Books.

Swartzman, Daniel, Richard A. Liroff, and Kevin G. Croke, eds. 1982. *Cost-Benefit Analysis and Environmental Regulations: Politics, Ethics, and Methods*. Washington: Conservation Foundation.

Switzer, Jacqueline Vaughn. 1994. *Environmental Politics: Domestic and Global Dimensions*. New York: St. Martin's Press.

Taubes, Gary. 1993. "The Ozone Backlash." *Science* 260 (June 11): 1580–83.

Tear, Timothy H., J. Michael Scott, Patricia H. Hayward, and Brad Griffith. 1993. "Status and Prospects for Success of the Endangered Species Act: A Look at Recovery Plans." *Science* 262 (November 12): 976–77.

Temples, James R. 1980. "The Politics of Nuclear Power: A Subgovernment in Transition." *Political Science Quarterly* 95 (summer): 239–60.

Terry, Sara. 1993. "Drinking Water Comes to a Boil." *New York Times Magazine* (September 26): 42–48, 62–65.

Tietenberg, Tom. 1994. *Environmental Economics and Policy*. New York: Harper-Collins.

Tobin, Richard J. 1990. *The Expendable Future: U.S. Politics and the Protection of Biological Diversity*. Durham: Duke University Press.

———. 1992. "Environmental Protection and the New Federalism: A Longitudinal Analysis of State Perceptions." *Publius* 22 (winter): 93–107.

———. 1994. "Environment, Population, and Economic Development." In Vig and Kraft, *Environmental Policy in the 1990s*.

Tong, Rosemarie. 1986. *Ethics in Policy Analysis*. Englewood Cliffs, N.J.: Prentice-Hall.

Underwood, Joanna D. 1993. "Going Green for Profit." *EPA Journal* (July–September): 9–13.

United Nations. 1993. *Agenda 21: The United Nations Programme of Action from Rio*. New York: United Nations.

————. 1994. *World Population Prospects: The 1994 Revision.* New York: United Nations Population Division.

U.S. Congress. 1988. *Environmental Federalism: Allocating Responsibilities for Environmental Protection.* Washington: Congressional Budget Office, September.

————. 1994. *The Total Costs of Cleaning Up Nonfederal Superfund Sites.* Washington: Congressional Budget Office, January.

U.S. Department of Energy. 1993. *Energy Facts 1992.* Washington: Energy Information Administration, DOE/EIA-0469(92), October 29.

U.S. Environmental Protection Agency. 1987a. "List of Committees and Subcommittees of Interest to EPA." Washington: EPA, Office of Legislative Analysis, June 16.

————. 1987b. "Congressional Hearings Held, 1984, 1985, 1986." Washington: EPA, Office of Legislative Analysis.

————. 1990a. *Reducing Risk: Setting Priorities and Strategies for Environmental Protection.* Washington: EPA, Science Advisory Board, SAB-EC-90-021, September.

————. 1990b. *Environmental Investments: The Cost of a Clean Environment: Report of the Administrator of the Environmental Protection Agency to the Congress of the United States.* Washington: EPA, EPA 230-11-90-083, November.

————. 1992a. *Framework for Ecological Risk Assessment.* Washington: EPA, Office of Research and Development, February.

————. 1992b. *Environmental Equity: Reducing Risk for All Communities,* 2 vols. Washington: EPA, Office of Policy, Planning, and Evaluation, EPA-230-R-92-008A, June.

————. 1993. *National Air Quality and Emissions Trends Report, 1992.* Research Triangle Park, N.C.: EPA Office of Air Quality Planning and Standards, EPA 454/R-93-031, October.

U.S. General Accounting Office. 1992. *Environmental Protection Issues, Transition Series.* Washington: Government Printing Office, GAO/OCG-93-16TR, December.

————. 1993a. *Electricity Supply: Efforts Under Way to Develop Solar and Wind Energy.* Washington: Government Printing Office, GAO/RCED-93-118, April.

————. 1993b. *Air Pollution: Impact of White House Entities on Two Clean Air Rules.* Washington: Government Printing Office, GAO/RCED-93-24, May 6.

————. 1993d. "Management Issues Facing the Environmental Protection Agency." Washington: GAO/T-RCED-93-26, March 29.

————. 1994a. *Drinking Water: Stronger Efforts Essential for Small Communities to Comply with Standards.* Washington: Government Printing Office, GAO/RCED-94-40, March.

————. 1994b. *Water Subsidies: Impact of Higher Irrigation Rates on Central Valley Project Farmers.* Washington: Government Printing Office, GAO/RCED-94-8, April.

U.S. Office of Technology Assessment. 1988. *Are We Cleaning Up?: 10 Superfund Case Studies.* Washington: Government Printing Office, June.

————. 1989. *Coming Clean: Superfund Problems Can Be Solved.* Washington: Government Printing Office.

————. 1991a. *Complex Cleanup: The Environmental Legacy of Nuclear Weapons Production, Summary,* OTA-0-485. Washington: Government Printing Office, February.

————. 1991b. *Improving Automobile Fuel Economy: New Standards, New Approaches, Summary,* OTA-E-508. Washington: Government Printing Office, October.

————. 1992. *Green Products by Design: Choices for a Cleaner Environment.* Washington: Government Printing Office.

Van Houtven, George L., and Maureen L. Cropper. 1994. "When Is a Life Too Costly to Save? The Evidence from Environmental Regulations." *Resources* 114 (winter): 6–10.

Vig, Norman J. 1994. "Presidential Leadership and the Environment: From Reagan and Bush to Clinton." In Vig and Kraft, *Environmental Policy in the 1990s.*

Vig, Norman J., and Michael E. Kraft, eds. 1984. *Environmental Policy in the 1980s: Reagan's New Agenda.* Washington: CQ Press.

———. 1990. *Environmental Policy in the 1990s: Toward a New Agenda.* Washington: CQ Press.

———. 1994. *Environmental Policy in the 1990s: Toward a New Agenda,* 2nd ed.. Washington: CQ Press.

Wald, Matthew L. 1992. "U.S. Finds Energy Industry Subsidies Are Small." *New York Times* (December 14):C2.

———. 1993. "After 20 years, America's Foot Is Still on the Gas." *New York Times* (October 17):E4.

———. 1994a. "First Prosecution Under Ozone-Protection Law." *New York Times* (May 13):A8.

———. 1994b. "Regulator Fears Nuclear Plants Will Scrimp." *New York Times* (September 9):A10.

———. 1994c. "U.S. to Aid Big 3 in Cleaner-Car Research." *New York Times* (October 19):C1, 2.

Wandesforde-Smith, Geoffrey. 1989. "Environmental Impact Assessment, Entrepreneurship, and Policy Change." In Bartlett, *Policy Through Impact Assessment.*

Washington Post. 1994. "Plan to Cut Emissions of Greenhouse Gases to '90 Levels Called Unattainable." *Washington Post* (September 3).

Wengert, Norman. 1994. "Land Use Policy." In *Encyclopedia of Policy Studies,* 2nd ed, edited by Stuart S. Nagel. New York: Marcel Dekker.

Wenner, Lettie M. 1994. "Environmental Policy in the Courts." In Vig and Kraft, *Environmental Policy in the 1990s.*

Whitaker, John C. 1976. *Striking a Balance: Environment and Natural Resources Pol-icy in the Nixon-Ford Years.* Washington: American Enterprise Institute.

Wildavsky, Aaron. 1988. *Searching for Safety.* New Brunswick, N.J.: Transaction Books.

Wilson, Edward O. 1990. "Threats to Biodiversity." In *Managing Planet Earth.* New York: W. H. Freeman.

Wilson, James Q. 1980. "The Politics of Regulation." In *The Politics of Regulation,* edited by James Q. Wilson. New York: Basic Books.

Wines, Michael. 1993. "Tax's Demise Illustrates First Rule of Lobbying: Work, Work, Work." *New York Times* (June 14):1, A11.

Wood, B. Dan. 1988. "Principals, Bureaucrats, and Responsiveness in Clean Air Enforcements." *American Political Science Review* 82: 213–34.

Wood, B. Dan, and Richard Waterman. 1991. "The Dynamics of Political Control of the Bureaucracy." *American Political Science Review* 85: 801–28.

World Commission on Environment and Development, 1987. *Our Common Future.* New York: Oxford.

World Resources Institute. 1992. *The 1992 Environmental Almanac.* Boston: Houghton Mifflin.

World Wildlife Fund. 1994. "Old-Growth Forests, Ecosystem Management, and Option 9." *Conservation Issues* 1 (May–June): 5–8.

Yaffee, Steven Lewis. 1994. *The Wisdom of the Spotted Owl: Policy Lessons for a New Century.* Washington: Island Press.

Yosie, Terry F. 1993. "The EPA Science Advisory Board: A Case Study in Institutional History and Public Policy." *Environmental Science and Technology* 27: 1476–81.

Young, John. 1994. "Using Computers for the Environment." In *State of the World 1994,* edited by Lester R. Brown et al. New York: W. W. Norton.

ABBREVIATIONS

AEC	Atomic Energy Commission
ANWR	Arctic National Wildlife Refuge
BAT	Best Available Technology
BOD	Biological Oxygen Demand
BLM	Bureau of Land Management
CAA	Clean Air Act
CAFE	Corporate Average Fuel Economy
CBO	Congressional Budget Office
CEQ	Council on Environmental Quality
CERCLA	Comprehensive Environmental Response, Compensation, and Liability Act
CFCs	Chlorofluorocarbons
CWA	Clean Water Act
DOD	Department of Defense
DOE	Department of Energy
EA	Environmental Assessment
EDF	Environmental Defense Fund
EIS	Environmental Impact Statement
EMAP	Environmental Monitoring and Assessment Program
EO	Executive Order
EPA	Environmental Protection Agency
EPCRA	Emergency Planning and Community Right to Know Act
ESA	Endangered Species Act
ETS	Environmental Tobacco Smoke
FDA	Food and Drug Administration
FIFRA	Federal Insecticide, Fungicide, and Rodenticide Act
FLPMA	Federal Land Policy and Management Act
FTC	Federal Trade Commission
FWPCA	Federal Water Pollution Control Act
FWS	Fish and Wildlife Service
GAO	General Accounting Office
GDP	Gross Domestic Product
GNP	Gross National Product
HSWA	Hazardous and Solid Waste Amendments
IPCC	Intergovernmental Panel on Climate Change
LCV	League of Conservation Voters
NAAQS	National Ambient Air Quality Standards
NAPAP	National Acid Precipitation Assessment Program

NEPA	National Environmental Policy Act	PSD	Prevention of Significant Deterioration
NES	National Energy Strategy	PURPA	Public Utilities Regulatory Policy Act
NFMA	National Forest Management Act	R&D	Research and Development
NIMBY	Not In My Back Yard	RARE	Roadless Areas Review and Evaluation
NOAA	National Oceanic and Atmospheric Administration	RCRA	Resource Conservation and Recovery Act
NPDES	National Pollution Discharge Elimination System	RIA	Regulatory Impact Analysis
NPL	National Priorities List	ROD	Record of Decision
NPS	National Park System	SARA	Superfund Amendments and Reauthorization Act
NRC	Nuclear Regulatory Commission	SAB	Science Advisory Board
NRDC	Natural Resources Defense Council	SDWA	Safe Drinking Water Act
		SIP	State Implementation Plan
NPR	National Performance Review	SMCRA	Surface Mining Control and Reclamation Act
NWF	National Wildlife Federation		
NWPA	Nuclear Waste Policy Act	SWDA	Solid Waste Disposal Act
NWPS	National Wilderness Preservation System	TRI	Toxics Release Inventory
		TSCA	Toxic Substances Control Act
OCS	Outer Continental Shelf	TVA	Tennessee Valley Authority
OIRA	Office of Information and Regulatory Affairs	UNCED	United Nations Conference on Environment and Development (Earth Summit)
OMB	Office of Management and Budget	UNEP	United Nations Environment Programme
OPEC	Organization of Petroleum Exporting Countries	UNFPA	United Nations Population Fund (formerly United Nations Fund for Population Activities)
OSHA	Occupational Safety and Health Administration		
OTA	Office of Technology Assessment	USDA	U.S. Department of Agriculture
PCBs	Polychlorinated Biphenyls	USTs	Underground Storage Tanks
ppm	parts per million	VOCs	Volatile Organic Compounds

INDEX